Endorsements for "Darwin's Replacement

"The need for a super-intelligent force to create and sustain living things is well set out and without question. I also appreciated all the research that was done to demonstrate the historical importance and recognition of God in the nation lives of the four nations you selected."

-- George Matzko, PhD

"*Darwin's Replacement* is a well written review of the enormous complexity of all life written from the perspective of the molecular foundation. The complexity is more than amazing. Most people enjoy learning about amazing feats, and thus the popularity of 'Ripley's Believe It or Not' and similar books. Our body and its complexity is familiar and works so well for most of us that we often take it for granted. Mr. Rogers' book helps us to realize that we are all walking miracles, and understanding how it functions is both awe inspiring and helps us to appreciate the body we all live in while on this earth."

-- Jerry Bergman, PhD

"*Darwin's Replacement* provides credible assessment of the weaknesses within mainstream evolution theory, and proposes a reasonable, evidence based alternative for the origin and development of life."

-- Nicholas Comninellis, MD MPH

"ATOMIC BIOLOGY promises to restore the true foundations of science back to the realm of observation. Current forays into metaphysical speculation and presupposition by 'experts' seem to have caused division, confusion, and misunderstanding in the scientific conversation."

-- Jack Taylor, PhD

"EARLY REVIEWS for DARWIN'S REPLACEMENT show the findings in the book being applauded as well-researched, timely, and logical." -- Keri Vermeulen, Editor, The Light Magazine

"My recommendation would be to rework the book as a series of 1-2 page studies for adult Sunday school classes and/or Christian High School classes. It could be useful as an introduction to the topics, in that format."

-- David Snoke, PhD

DARWIN'S REPLACEMENT

"Your book demonstrates the amazing complexity of life, starting with even the simplest cell, and the numerous conditions needed to sustain life. That all this could be the result of blind, random evolution is highly implausible, and statistically virtually impossible. Hence, as your book concludes, this points to a super-intelligent Creator. Your book also notes that the USA, the UK, Canada, and Australia were all founded on submission to the Christian God, and urges those countries return to acknowledging God, also in science classrooms.

I heartily agree with all this." -- John Byl, PhD

"There are only two possibilities for the existence of life: accidental or purposeful. Using science and mathematics, Atomic Biology proves beyond a shadow of doubt that life cannot be accidental. Then the book shows that the only being capable of the creation of life and its orchestrated maintenance is the historical Omniscient, Omnipotent, Omnipresent Triune God of the Bible and our Nation."

Sharon E. Cargo, DVM

"Hi, (Reality R&D) I recently bought the book and am reading it slowly, and out loud to myself so that it sticks, but can I say that when I heard Mr. Rogers on Vision I knew that this was a book I'd been waiting for a very long time. I honestly can't put into words how exciting this is for me to finally have something I can refer to when discussing creation and not sound like a loony."

-- Linda Houston, Australia

Flags of Four of the Nations Where God Is Part of Government

DARWIN'S REPLACEMENT

A primer introducing the Godly life-science of Atomic Biology

OlgaLyubkina/Shutterstock.com

See how our Creator faithfully and reliably constructs enough food every day to feed us and our seven billion neighbors, plus all the other creatures and plants. Then He makes our cells from our food. This book will show how we are made from the "dust" through His phenomenal and careful two-step process. Hopefully you will see the simple science of it all to be personally fascinating.

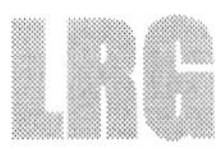

LIFETIME REFERENCE GUIDES INC.
P.O.Box 51613 RPO Park Royal
West Vancouver, BC, Canada V7T 2X9

www.lifetimereferenceguides.com

Copyright © 2017 by Lifetime Reference Guides Inc.

First Edition – 2017

Cover design by ArneeonMedia.com with Shutterstock images.

All rights reserved.

No part of this publication may be reproduced in any form by any means, electronic or mechanical, including photocopying, recording, information browsing, storage, or any retrieval system, without specific written permission from the publisher.

ISBNs
978-0-9940786-4-3 (Hardcover)
978-0-9940786-6-7 (Paperback)
978-0-9940786-5-0 (eBook)

Science
Education
Government

Contents

Part II: Teaching The "Why's" of God's Inclusion In Our Governments, and Why This Ties In with the Science of Atomic Biology

Dedication

I (also) have a dream, and our Creator could have the same one: that all people in science and education would agree and teach that super-intelligence, immense care, and super-abilities, far beyond those of mankind, are essential for designing, building, and sustaining all living cell-parts, cells, and entities, including us.

Our Creator would get the credits due, and our students would get accurate knowledge of their caring life-cause.

- Tom Rogers

This book provides simple, significant, scientific solutions to show why Evolution is NOT the cause of life, but, as the majority of citizens and the government founders have believed, God IS.

Prologue

In 1987, when I finally "saw the light," it did not take long for my curiosity to be piqued regarding the Creator whom I had mostly ignored for the previous four decades.

Having education in three universities and two institutes and a work background in engineering, research, construction, and manufacturing, I had to find out how creation by a Creator works.

My awards were in biology, chemistry, physics, and economics, but the majority of my science education has come through research during the last thirty years.

Remembering as far back as my grade 5 science class and ever since, that material things are made of atoms, that seemed like a good place to start.

The surprising thing that bothered me most was that I could not find any books that detailed the assembly of atoms into cells. My thoughts were, "You've got to be kidding! This is so basic! Somebody must have figured this out!"

It seemed that the assembly books were not to be found. I looked at Darwin's original Theory of Evolution plus the modifications that followed, but had learned enough to know there is an enormous amount of intelligence necessary for building living cells and entities. All the varied theories of evolution claim there is no need for intelligence. Knowing this to be a mistake, it was easy to see that evolution was not the cause of life.

Eventually, after researching some examples, it became very clear just how fantastic the cause of life really is; e.g. after piecing numbers together from three other scientists, the result shows that over 4900 quadrillion right atoms per second are selected from our digestive system and precisely assembled for each one of us, just for replacing our worn-out red blood cells. (See development later).

As little was written on the details of cell construction, my calling became clear to put these research results into writing. I had joined several science associations but found in speaking to a number of members that none of them had really thought through the details of the brilliant physical work necessary for selecting and assembling atoms to build our cell-parts, cells and us.

More and more pieces of the puzzle were provided and my purpose made clear.

The BIG question is: IF it can be shown that *Intelligent Physical Works* are essential for building living cells and entities, does this *Falsify Darwinisms and Macro-evolution as the cause of life?*

Many won't like the answer but it has become an obvious, "Yes."

In 2016, Jean-Pierre Sauvage, Sir J. Fraser Stoddard, and Bernard L. Feringa won the Nobel Prize in Chemistry for their development of some simple molecular machines. Their intelligent work made a significant step forward in 1983 when Sauvage found a way to link two special molecules together, but it took more intelligent input from Stoddard and Ferringa over the next thirty years to finally develop a few relatively simplistic molecular machines.

Now please note this: Each of us have about 100 trillion cells, and each cell has several molecular machines built-in which are almost infinitely more complex than the best built by scientists. This obviously requires more intelligence and dexterity than mankind has, therefore we can conclude that the builder has super-intelligence, super-dexterity, super-speed, and super-care for us.

The "super" prefix means "far beyond the scientific capabilities of mankind."

All this to begin to show why evolution alone, by definition having no intelligence to use, cannot build any living cells or cause life.

The main purpose of this book is to show many ways super-intelligence is essential to create living entities, including us.

This is not just a primer, it is a launch pad for the major project of revealing solid reasons, both scientific and historical, why the God of our Governments, should be invited back to our public schools, colleges, and universities, especially into our science classes.

For those concerned about separating the church from the state, we make it clear that God is definitely not "the church," as we will show later. He is, however, a significant part of some churches, just as He is a significant part of our governments, and this gives our students the inalienable right to be taught something about Him.

According to the most recent censuses, the majority of citizens in the democracies of the USA, the UK, Australia, and Canada, believe in their Creator who is also the God of their governments. We will provide many good reasons to be grateful for His phenomenal scientific work, care, and provision for all of us.

He is often asked to bless our nations, so let's give Him more opportunity to do that. If we continue to lock Him out of our public education systems, why would we expect Him to continue to bless us fully?

To a dangerous degree, we have become like frogs in the water-pot on the stove. We have allowed certain vested interests to convince our education leaders that God is not our Creator or cause and sustainer of our lives, and that evolution-only is to get the credit.

Generally, we tend to believe our scientists, because, after all, they are the ones whose careers are supposed to be focused on studying how things truly work. But what if there is major disagreement amongst scientists on crucial issues, as there is in the Creator vs Darwin debate? There is a huge misconception by some that Evolution is more scientific than God, but nothing could be further from the truth, as we will show.

After almost three decades of research into the super-intelligent works essential to build us from the "dust," our findings, at Reality Research & Development, with input from 20 PhDs, 9 DScs, 3 MDs, and other scholars, provide solid, knowledgeable, and logical reasons for dismissing evolution as the cause of life.

There are often minor changes between generations of species, but even these changes do not "just happen" by random chance or unguided process. All cells, including mutation cells, have to be physically constructed using super-intelligence for decisions and choices, atoms for building blocks, and super-dexterity for assembly of the right numbers of the right atoms to make our cells and us.

Darwin himself obviously believed in God as he mentioned our "Creator" several times in his books with statements like, *"... the Works of the Creator are (superior) to those of man."* [1]

This is so true. With all man's accumulated scientific knowledge and sophisticated equipment, we cannot come anywhere close to building even one of the tiny molecular machines essential for each living cell. So why would we expect that an unintelligent theoretical process, like evolution, could build whole cells and living entities?

For many in science today, there is an obnoxious (perhaps illegitimate) pressure to "toe the line" of "evolution only" as the taught, used, and allowed "cause-of-life" in our public education classrooms. As shown in the movie, "Expelled! No Intelligence Allowed!"[2] severe reprimand and dismissal are often the ugly penalties executed upon professors or teachers who dare to suggest, especially to students, that there may be an intelligent design and cause of life.

This exclusion attitude is totally *anti-science* and has to be stopped. (Education leaders, please note). Science is supposed to "Go Where the Evidence Leads" without detrimental (perhaps illegal) restrictions.

There are now lists of thousands of scientists in less threatened positions, who are publicly declaring their skepticism and/or total disbelief regarding Darwinian evolution as the cause of life. There are also many other more threatened scholars, including educators, who dare not go public yet, with their skepticism or disbelief regarding evolution, because of the threats of penalties.

Some of the lists include Dr. Jerry Bergman's lists of "Darwin Skeptics," the Discovery Institute's list called "A Scientific Dissent from Darwinism", and more than four other lists, totaling some 2.5 million professional scholars in the U.S. alone.

This primer has two parts which some think should be in separate books. However, the major purpose is to show why the scientific Creator God of our nations should be made known to all of our students, for scientific reasons:

- **Part I** provides the founding details of the Godly life-science we are calling "Atomic Biology". We could call it a 'new' science, but God has probably been using it since the beginning, along with gravity, electricity, magnetism, physics, chemistry, math, and so on.

  (It is becoming more apparent that most sciences are developed to study what God is already doing, Biomimicry being a prime example).

- **Part II** reminds us of many ways God is part of our governments in the USA, the UK, Australia, and Canada. This fact gives our students and other citizens, the inalienable right to be taught the reasons 'Why' He is, and always has been, significant to us and our nations. This book will help to clarify some of the great benefits in understanding His immense *scientific care* for each of us.

And just to be clear, the God of our governments in these four nations, is the triune Creator God of the Holy Bible, not any other.

As mentioned above, He is not to be confused with "the Church" because a Church can be a building, or a group of people, who are Satanists, Scientologists, New Agers, Catholics, Protestants, Mormons, Jehovah's Witnesses, cults, etc., etc., etc. God is definitely NOT "the Church."

He is a significant part of some churches, just as He is a significant part of our governments as shown by His inclusion in National Holidays (Easter, Thanksgiving, Christmas), in Anthems ("God Save The Queen", "God Keep Our Land Glorious and Free", "God Bless America"), in Declarations, Pledges of Allegiance, Constitutions, Oaths of Office, on Currencies ("In God We Trust", "Dei Gratia Regina"), on Public Buildings, War Memorials, in National Prayers, and so on.

The concept of separating "The Church" from "The State" was purposefully legislated in the USA so the State cannot tell Religions what to do (other than obeying the laws), and Religions cannot tell the State what to do.

The God of our nations is not a religion, although He is a significant part of some religions, just as He is a significant part of our governments.

Belief in Evolution as the cause of life, knowing how

brilliant and complex life is, requires immense Faith. This makes Evolution as much of a Religion as any other Belief requiring Faith.

There are several purposes for this primer which can be beneficial for citizens, sciences, students, and governments.

For citizens, bringing the highly-respected God of our nations back to our classrooms, especially for scientific reasons, can bring a renewed guidance for students of any age, in finding the best that life has to offer. Citizens could be taught the reasons for His respect and role in governments, the joy available through His teachings to help others, the benefits to all, of strong moral character, respect for authority where earned, and the beneficial understanding of choices and consequences. In addition are the many caring scientific works He performs in creating, sustaining, maintaining, and repairing each of us, our foods, our pets, and the other living plants and creatures around us.

We outline how His guidance has made these four nations so desirable and attractive to people from other nations where His advice is not heeded.

The major concern now, however, is that our governments are, more and more to our detriment, ignoring God's wise advice.

He does not force anyone to do anything, but His advice is certainly worth following.

For the sciences and students, we introduce a new branch of biology for determining the super-intelligent works necessary for building, sustaining, maintaining, and repairing living cells and entities, including us. The name we have adopted for this science is "Atomic Biology", as it goes deeper than molecular biology, down to where the right numbers of the right atoms have to be found in available resources and precisely assembled into our cells and us. Applications will be found in the near future, for improving nutrition, medicine, agriculture, aquaculture, and other related sciences and social studies.

In addition, for students, citizens, and governments, are recommendations to seek real wisdom regarding choices and consequences. Best choices obviously lead to best results.

For governments, we provide several reminders of the benefits of the God-given principles that have made our four nations so attractive to peoples from nations where His advice is not heeded. We also provide warnings of the dangers of moving away from those principles that have kept us strong, safe, and blessed as societies. Immigrants have left less principled governments to come to our lands, and we have welcomed them with some stipulations. However, we are foolish to alter our standards to what they left.

We are recommending that Darwinisms be moved to the history department as falsified theories for the cause of life. Evolution alone is not an intelligent force and cannot even start finding the right numbers of the right atoms in available sources to build any living cell, as, by definition, it has no intelligence to use. False information is a dull tool to provide to those we want to ultimately find solutions to our various problems.

For the technical basics, we show seven principles and eighteen essentials for life that require God's super-intelligent works. When any of these points are verified, evolution is disqualified as the cause of life.

Darwin's Replacement is a culmination of almost three decades of research including significant input from 20 PhDs, 3 MDs, 9 DScs, 3 Mathematicians, 2 MScs, and 8 Independent Researchers.

Our driving belief is this: ***Citizens, including our students, have the inalienable right to be taught Why God is a highly recognized part of Their Government and what phenomenal scientific work and care He is providing for each one of us, every second of every day.***

-Tom Rogers, President
Reality Research & Development Inc.

Glossary

Normally, the glossary would be in the back of the book; however, some of our definitions of old terms are quite new. Without having the definitions used by the authors, you as the reader, would not receive clarity for the messages herein.

Although some of the terms may seem technical, you will find that this is not "rocket science". We believe you will find it much more personally beneficial than rocket science. Many of these terms are interrelated.

Atomic Biology
This is a name that we, at Reality Research & Development Inc., are giving to the study of the enormous amount of intelligent physical work essential for finding, sorting, selecting, counting, grasping, and precisely placing and fastening all the right numbers of the right atoms required for constructing, sustaining, growing, maintaining, and repairing living cells and entities. This is a Godly life-science.

(With each discovery in science, it is becoming more apparent that most sciences are studies of God's super-intelligent, enormous, and careful works).

Atoms
They are commonly referred to as the "building blocks of the universe". Virtually all atoms are comprised of a nucleus made up of protons and neutrons with electrons being moved *perpetually* in orbits around the nucleus at controlled speed so as not to fly out of orbit by being moved too quickly, nor to implode into the nucleus by being moved too slowly. Atoms have no *internal* means of self-

directed movement, no legs, fins, wings, muscles, or brains, and therefore must rely upon a super-intelligent *external force* to move them into their precise position in a cell.

Biological Construction

This is the super-intelligent process of building living cells and entities by finding, sorting, selecting, counting, grasping, and precisely assembling and fastening all the right numbers of the right atoms from available sources. This is essential to build each complex part of every living cell, along with adding the necessary 'breath-of-life' to each cell, precisely programming its DNA, RNA, etc. to make it specifically functional, and assembling various specialized cells into the particular living entity desired by the builder.

Breath-of-Life

The super-intelligent basic necessity for life in every living cell, without which no cell lives or functions.

Common Descent

An evolutionary concept stating that similarities between different kinds of plants and creatures indicate a common ancestor.

However, a better understanding of the intelligent design, construction, sustenance, and maintenance work required for living entities reveals the necessity for a common designer, builder, and maintainer.

Creation

The intelligent work necessary for the design, construction, sustenance, growth, maintenance, and repair of living entities, by a super-intelligent force using non-living elemental atoms.

Decisions and Choices for Cell Construction
A key factor in the work of assembling atoms into cell parts is deciding on the sequencing of the assembly of the right numbers of the right atoms, then choosing the right atoms from amongst the wrong ones and assembling them into each part of each cell in proper sequence with precise placement and fastening of each atom. It takes super-intelligence to do this work and knowledge of how each part of each different cell has to function on completion. Think of the intelligent work necessary in assembling atoms for building the DNA molecules in each cell, for example.

Design
The super-intelligent process of planning the appearance and functions of each cell for each living entity, planning its source of the right atoms for its building materials, planning the right numbers of the right atoms to be selected from amongst millions of unsuitable atoms, planning the DNA and RNA programs for the functions of cell parts, and designing the molecular machines to be built within each different cell, etc.

Energy
The super-intelligent force supplied consistently, carefully, and constantly to provide intelligently controlled perpetual motion to every electron in every atom.

Evolution
Because this topic is so controversial, it is important to any discussion of it that the individuals agree on which definition of 'evolution' they wish to discuss.

Virtually every interested individual can agree that there is the kind of minor or 'micro' evolution which means minor changes from generation to generation in hair color, eye color, body size, and even skin color with children of mixed-race marriages. What is not usually considered is that all the cells, including the different-styled ones from generation to generation, and mutation cells, all have to be intelligently constructed of selected atoms.

Humans only have human children; monkeys only have monkey children, etc. One kind of creature does not give birth to another kind of creature, even after long periods of time.

For this book, the evolution we are referring to is the kind currently being taught in the majority of our government-funded grade schools, colleges, and universities, i.e. Darwinism, Neo-Darwinism, or macroevolution. These are theoretical explanations for the cause of life and species whereby all species originated and descended from a single organism that was formed by a chance assembly of atoms billions of years ago. There is no guidance, purpose, or intelligent work involved with these theoretical processes.

The original chance organism would have needed the highly complex ability to live, to function, to find nourishment and digest that to sustain its life, to reproduce, to remember its good qualities and pass them on with improvements, and to change into all the other kinds of plants and animals including, eventually, monkeys or their ancestors, ascending into mankind, and to do all of this with no intelligent help, through a totally unguided process. Because of the extreme improbability, now known, for such a complex original life form to just "happen," even many evolutionists, as well as former evolutionists, are moving further away from the concept.

Evolution can be defined as a form of religion, as it is a belief system requiring enormous Faith that a totally unintelligent process could assemble anything as phenomenally complex as living cells and entities.

God

The name given by English speaking people and governments to the super-intelligent force and entity who creates and sustains all other living entities. He is like eternity and infinity in that there is no known beginning or end for Him. His intelligently controlled energy is required to consistently move every electron in every atom in every element perpetually at the right speed. There is no other known source of this essential and eternal supply of controlled energy for the perpetual motion of electrons. This partially explains His omnipresence, and omniscience as He has a presence in all the atoms in all living cells. He is the brilliant designer, builder, sustainer, maintainer and repairer of every living cell in every living entity including those in the food items necessary for our life. He is the provider of the super-intelligent 'breath-of-life' required by each cell of every living entity in order for it to live and function; no cell lives without it. He is the programmer of all DNA and RNA, and the builder of all the molecular machines and other parts for every cell. His phenomenal works, capabilities, and care seem unlimited except by His own will. This earns Him the description as omnipotent.

God is not to be confused with the "Church" because the Church can be a building or a group of people who are Satanists, Scientologists, Catholics, Protestants, Mormons, Jehovah's Witnesses, New Agers, or cults, all of whom cannot create any living thing.

Separation of Church and State is a great plan to prevent one from telling the other what to do within the law. Just remember that God is not the Church, although He is a major part of some churches, just as He is a major part of our Governments. His advice is worth using.

Intelligence

The ability to calculate the requirements for solutions to objectives and to provide the means to fulfill these requirements. Intelligence has to be super-intelligently high when it applies to the calculations, provisions, and works necessary to design and construct *all parts necessary for the life functions in every living entity.*

Molecular Machines

Of particular interest are the amazing tiny machines constructed within each of our cells to perform various functions.

The 2016 Nobel Prize for Chemistry was given to three scientists, Sauvage, Stoddart, and Feringa, who had cooperatively researched and developed a few simple molecular machines over the previous four decades.

Even though these molecular machines required so much scientific research, human intelligence, sophisticated equipment, and much time, they are extremely simple in comparison to any of the extraordinarily complex molecular machines like kinesin, ribosomes, myosin, proteosomes, spliceosomes, or cohesin, that God has to build for us within our cells in very short periods of time.

Maintenance

The super-intelligent work involved in the maintaining and supporting of cells for life extension in every living entity. This includes the delivery of the right form and amount of energy required for each cell's warmth and function, the delivery of the right type and amount of nourishment for its life, removing its waste products, and disposing of that through the entity's waste disposal system.

Natural

An adjective to describe a temporary entity having limited lifespan and capability.

Natural Selection

A term used in Darwinism/macroevolution to describe the survival of the fittest organisms. Its logical concept is that those living entities with characteristics that weaken their capacity to survive (such as the inability to escape predators or obtain nutrients) will have less chance of reproducing and passing on their characteristics to succeeding generations.

Natural selection is also the theoretical "mechanism" for producing and maintaining life, however, it is not an intelligent force and does not have the required capabilities to do all of the necessary *intelligent work* for constructing cells. These missing capabilities include: the vision to find and count all the right numbers of the right atoms in available sources; the means of grasping atoms; the speed and dexterity for precisely placing the right numbers of all the right atoms together to build any cell; the intelligence to program DNA and RNA; and the power to provide the necessary 'breath-of-life' without which no cell lives or functions. In short, natural selection/evolution does not have the necessary capabilities to construct, sustain, grow, maintain, and repair even one cell. There are all these many super-intelligent works necessary to create living entities which natural selection/evolution *cannot perform.* Many semi-committed evolutionists and former evolutionists now acknowledge this fact.

Parameters of Possibility (for natural selection)

A maximum or limit to the capacity for accomplishment of objectives without super-intelligent help. For example, if you have a large bowl full of marbles of ten different colors, how many groups of ten different colors do you think you could sort, select and pick out to place in a circular pattern in one second? Maybe 2? Not a very large number. This is while using your intelligence, eyesight, arms and hands – there are parameters and limits to the number you can handle. Now, let us suppose that you have no brain, no intelligence, no vision, and no appendage for grasping atoms, as with the 'natural selection' theorized by Darwinism and macroevolution. Then how many could you sort, select, pick out, and place in a circular pattern in one second? None? The point is that even with the best brain and intelligence of any species on Earth, there is a limit or parameter to the number of items we can sort, select, grasp and place in a pattern in one second.

Now compare this to the work necessary just for the replacement of all our red blood cells about every 120 days. For just one average-sized human adult, *over forty-nine hundred quadrillion (4,900,000,000,000,000) right atoms per second* have to be sorted from his or her digesting food, then selected, counted, grasped, assembled into new red blood cells, and delivered into his or her blood stream (see Chapter 6 References). The number is huge for children as well. This is just a portion of the enormous amount of work performed constantly and reliably for virtually every person every second of every day. This requires super-intelligent speed, precision, intelligence, dexterity, reliability, endurance, and care. *These are essential capabilities for causing life, which 'evolution' does not have.*

Perpetual Motion

Regarding electrons in atoms, this is their consistently ongoing movement using the constantly controlled energy supplied by the super-intelligent force we call 'God'. To our understanding, it applies to the movement of electrons virtually forever in virtually all atoms. It is a super-intelligent capability involving a carefully controlled energy supply, intelligence, precision, consistence, and work. Perpetual motion is not possible without super-intelligent help.

Selection

The deliberate and intelligent choosing of the right numbers of the right atoms as part of the process needed to build, sustain, grow, maintain, and repair living entities.

Super-intelligent

An adjective to describe the perpetual force having abilities that immensely surpass human intelligence, speed, dexterity, capability, and endurance. See the definition of *God.* (A case can be made for the existence of an opposing force known as "Satan" who tempts people into much trouble, but we are dealing very little with that super-intelligent force in this book).

Super-intelligent Reliability

The consistently predictable, super-intelligent, Godly works provided by the designer, builder, sustainer, and maintainer of all living entities. A few examples we can rely upon Him for are:

- *gravity* to keep material items like us from flying off the face of the Earth as it spins upon its axis;
- *sunlight* for warmth, energy, and its part in the growth and maintenance of living entities;

- the *repositioning and healing effect of pharmaceutical atoms* ingested to help healing and to relieve pain;
- the *building* of Red Delicious apples when Red Delicious apple seeds are planted, tomatoes when tomato seeds are planted, carrots when carrot seeds are planted, etc.

All of these *and much more* are critical to life, yet because of God's amazing reliability and consistency, all these brilliant super-intelligent works and care are too often taken for granted.

Sustenance
The foods and beverages created by God for the nourishment of each and every one of His plants and creatures. From the available sources of the right numbers of the right atoms, He sorts, selects, grasps, and precisely assembles these foods and beverages for use by His creatures especially in developing energy for warmth and mobility of all muscles, for growth, for maintaining and repairing cells, for the phenomenal abilities to think, live, function, etc.

Works
The super-intelligent efforts which are *essential* in the work of designing, building, growing, sustaining, maintaining, and repairing each living entity, as the builder chooses.

Part I:
Introducing the Godly Life-Science of "Atomic Biology" as the Replacement for Darwinisms

Chapter 1

The Essentiality of a Super-Intelligent Force

Quotes from Charles Darwin:

"...the Works of the Creator are (superior) to those of man." [1]

"But just in proportion as this process of extermination has acted on an enormous scale, so must the numbers of intermediate varieties, which have formerly existed, be truly enormous. Why then is every geological formation and every stratum not full of such intermediate links? Geology assuredly does not reveal any such finely graduated organic chain; and this perhaps is the most obvious and serious objection which can be urged against the theory (of evolution)*."* [2]

"To suppose that the eye with all its inimitable contrivances for adjusting the focus to different distances, for admitting different amounts of light, and for the correction of spherical and chromatic aberration, could have been formed by natural selection, seems, I freely confess, absurd in the highest degree." [3]

"If it could be demonstrated that any complex organ could not possibly have been formed by numerous successive slight modifications, my theory would absolutely break down." [4]

Quotes from scientist I. L. Cohen:

"At that moment, when the RNA/DNA system became understood, the debate between Evolutionists and Creationists should have come to a screeching halt." [5]

(He is rightfully indicating that when the enormous and super-intelligent complexities of RNA and DNA Programming were discovered, the theory of evolution, which, by definition, has no intelligence to program with, should have been dropped right at that time).

"Any suppression which undermines and destroys that very foundation on which scientific methodology and research was erected, evolutionist or otherwise, cannot and must not be allowed to flourish. ...It is a confrontation between scientific objectivity and ingrained prejudice – between logic and emotion – between fact and fiction. ...In the final analysis, objective scientific analysis has to prevail – no matter what the final result is – no matter how many time-honored idols have to be discarded in the process....

It is not the duty of science to defend the theory of evolution, and stick by it to the bitter end – no matter what illogical and unsupported conclusions it offers... If in the process of impartial scientific logic, they find that creation by outside super-intelligence is the solution to our quandary, then let's cut the umbilical cord that tied us down to Darwin for such a long time. It is choking us and holding us back.

...Every single concept advanced by the theory of evolution (and amended thereafter) is imaginary as it is not supported by the scientifically established facts of microbiology, fossils, and

mathematical probability concepts. Darwin was wrong. ... The theory of evolution may be the worst mistake made in science." [6]

The Theory of Evolution is based upon the premise that since minor changes in a species can occur from generation to generation, therefore, over a long period of time, major changes may accumulate. This could lead to entirely new species that are the result of the fittest survivors being selected naturally while the less fit do not survive.

On the surface, this seems very logical and straight forward. There is (theoretically) no intelligent creation required as entities just reproduce naturally, and all living species descended from a tiny common ancestor according to Darwin's "Tree of Life."

However, there are several major essentials conveniently overlooked by the theory, some of which are now recognized even by remaining evolutionists and former evolutionists; e.g., the now understood phenomenal complexity which would have been required by the original living entity, and the awesome complexity in the construction of cells today.

We will show that even micro-evolution does not 'just happen' without the super-intelligent physical work required in finding, sorting, selecting, and precisely placing in sequence, all the right numbers of the right various elemental atoms from available sources. This work is essential to construct every complex part of every cell. Then the "Breath-of-Life" has to be added to these inanimate atoms to make the cell-parts function. When this is removed, the cell's life is over.

Cells also require highly complex and guided molecular machinery to be constructed within them. Ask the James Tour Group at Rice University how much intelligence it takes to design and build even the simplest molecular machine. You can also read about the intelligent physical work it took for Jean-Pierre Sauvage, Sir J. Fraser Stoddard, and Bernard L. Feringa to build simple molecular machines before winning the 2016 Nobel Prize in Chemistry.

In spite of the lengthy, costly, and complicated work needed to develop these machines, they are extremely simplistic compared to molecular machines constructed for our cells, including kinesin, ribosomes, myosin, proteasomes, spliceosomes, cohesin, etc. (See Chapter 4).

Evolution is not a force, and it cannot even start doing this super-intelligent cell construction work, as, by definition, it has no intelligence to use.

The Godly life-science of "Atomic Biology" we are introducing, includes Intelligent Design but goes far beyond, through the enormous amount of essential Intelligent Physical Works that must be performed in constructing, sustaining, maintaining, repairing and replacing the cells of creatures such as us.

Atomic Biology goes beyond Information to Decisions, Choices, and Precision Works.

It goes far beyond the Parameters of Possibility for Natural Selection and Random Mutation to explain the cause of Life.

It is our hope that we can one day combine all these concepts under one title for the science that truthfully details the brilliant, caring work of creating, sustaining, and maintaining each living entity.

Our focus at Reality R&D, is not on the unknown date that life began on Planet Earth, but on understanding the brilliant details necessary for living cells and entities to be assembled, sustained, maintained, repaired, and replaced, today.

A prime example of God's type of observable macro-evolution is the changing of a relatively unattractive *crawling* creature into a beautiful *flying* creature within 12 to 14 days, not millions of years. Such is His work in dissembling a crawling caterpillar down to an atomic soup in a chrysalis, then brilliantly reassembling most of the same atoms into a beautiful flying butterfly.

We have a name for this work, "metamorphosis", but the name does not do the essential, super-intelligent demolition and reconstruction work.

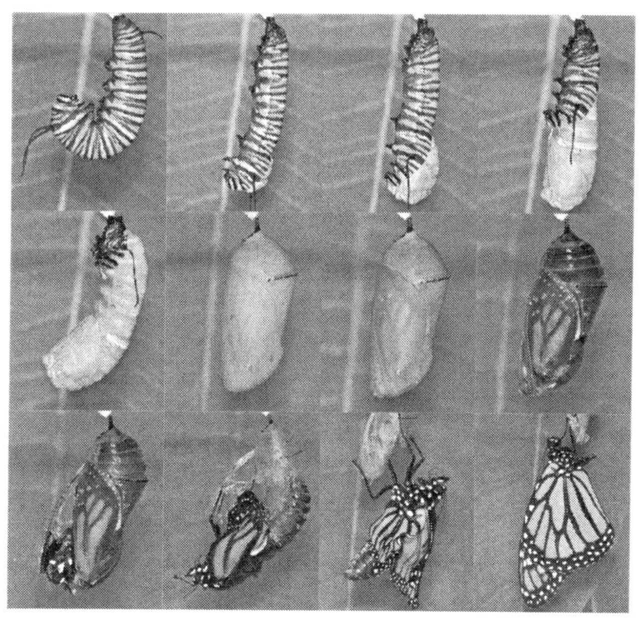

Stephen Russell Smith Photos/Shutterstock.com

Far more intelligent, awesome, and complex a task is building us humans from "the dust".

The concept for this is simple, but the work is virtually miraculous.

If you look at any group of cells in your body, e.g. skin cells, fingernail cells, foot cells, nose cells, you can ask, "From where did the atoms come for these cells?"

Answer: Where else than from the food we ate and a little from the air we breathed.

"And from where did those food atoms come?"

Answer: Where else than from the soil and moisture in gardens, fields, and orchards, and a little from the air.

So, this is why we know we are made from the dust: from atoms in the soil to our foods, and from many of the same atoms in our foods to our cells and us.

This is the easy and logical part. The "How" is the phenomenally complex part. It takes far more intelligence, dexterity, design, decisions, choices, speed, and precise sequential construction work, than scientists can come anywhere near. This is one of the reasons why Darwin was right when he said in *Origins, "... the works of the Creator are (superior) to those of man."* [1] even though he did not know all the details.

We will provide some of the basic details of these essential brilliant works within this book.

The more we learn about the marvelous molecular machines built into our cells to do the variety of jobs required therein, the more the case is strengthened, not for a Common Ancestor, but for a Common Designer, Builder, and Maintainer of all living entities.

This book will help to clarify some of the great benefits in understanding our Creator's immense care for each one of us.

As mentioned in the Introduction, just to be clear, the God of our governments in these four nations, is the triune God of the Bible, not any other. He is not to be confused with "the Church," as a church can be a building or a group of people who are Satanists, Scientologists, New Agers, Catholics, Protestants,

Mormons, Jehovah's Witnesses, cults, etc., etc., etc. God is definitely NOT "the Church".

God is, however, a significant part of some churches, just as He is a significant part of our governments as shown by His inclusion in National Holidays (Easter, Thanksgiving, Christmas), in Anthems ("God Save The Queen", "God Keep Our Land Glorious and Free", "God Bless America"), in Declarations, Pledges of Allegiance, Constitutions, Oaths of Office, on Currencies ("In God We Trust", "Dei Gratia Regina"), on Public Buildings, War Memorials, in National Prayers, and so on.

This is repeated because some readers skip the Introduction.

We should never lose our beneficial respect, acknowledgement, and appreciation for our Creator, Provider, and best Advisor.

The central theme for the basic science chapters herein, is to show a portion of the enormous amount of brilliant, caring, physical work involved in designing and constructing each unique one of us. We want to show the phenomenal design, intelligence, speed, dexterity, and reliable work and care required to find, sort, select, grasp, and precisely assemble the huge numbers of the right atoms to build each highly complex part of each of our individual cells. We want to show that there are approximately 100 trillion functioning cells that are magnificently made, placed, fastened, and amazingly hooked-up to our blood system for nourishment, waste removal, repairs, replacement, and temperature control for each one of us. Also, our cells are carefully hooked-up to our created miles and miles of nerve systems, and electrical systems for muscle actions, problem monitoring, and internal communication for all our senses and reactions. Our cells are wonderfully and carefully assembled to create each one of us as a uniquely special human being.

Similar super-intelligently brilliant physical work is performed for all creatures and plants, and we humans are given dominion over them all. May we have wisdom to do this well.

We must remember that atoms, of which every material thing is made, do not have legs, muscles, or brains. They cannot jump into their precise position in any cell because they do not have the internal means to do this. Of necessity, they require a *super-intelligent external force* with capabilities that go far, far, far beyond any unguided process like evolution.

This external force has to build our best foods using elements in the soil, air, and moisture in our fields, gardens, and orchards. Then, once we have eaten our foods, this force has to quickly sort out the right numbers of the right atoms, select and grasp onto them and fasten each atom into its precise place to make each highly complex part of each cell for our body. Then each newly created cell requires the super-intelligent "breath-of-life" to be added because atoms have no life of their own, only their given energy. Of course, this phenomenal force is our creator whom we call "God". He is always with us and within us.

This chapter could contain many volumes of detailed lists yet still miss huge numbers of the amazing works our Creator and Sustainer performs for us every second of every day.

Let's consider some of His phenomenal *work and care* for us humans and other marvelous creatures and plants He has made.

1. He Builds Our Best Foods

monticello/shutterstock.com

It does not take "rocket science" to understand that this essential work is performed for us on a daily basis. As soon as we recognize that material things including our foods are made of atoms and that atoms do not have legs *or any other internal means* to move themselves from their position in the soil, air, or moisture to their precise position in each cell for each morsel of our grown food, we can understand that *a super-intelligent external force is necessary*.

This applies not only to each cell of every tiny root, and to the body of each vegetable or fruit, but to the leaves as well. There are multi-millions of various cells in each morsel of food, and each cell requires millions of the right atoms to be precisely placed and fastened.

Cell construction, including the programming of DNA in each cell, does not happen by magic, magnetism, chemical reactions, electronics, or evolution. It requires a huge amount of super-intelligent, physical work and care.

The amount of food manufactured from the soil, air, and water around the world has to be sufficient to feed, without stopping, each one of us and more than seven billion global neighbors every day of every year. How enormous and reliable is that? Just think about this vast work and care for a moment.

This is the main reason why our national governments established a special holiday to give thanks to our Creator and Sustainer for all His caring work for us. It is called "Thanksgiving Day".

Some of us give thanks to this wonderful Provider, at every mealtime.

We could learn to share the foods better.

2. He Builds Our Babies and Our Bodies

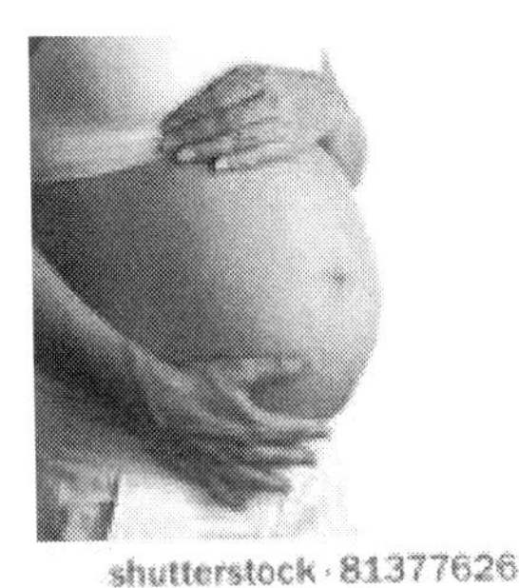

shutterstock · 81377626

Aaron Amat

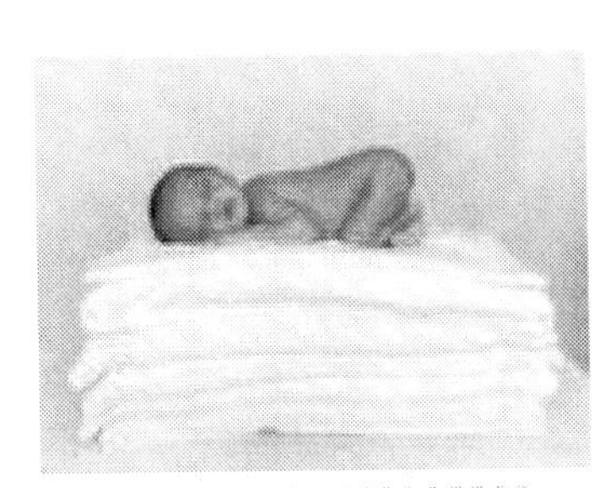

shutterstock · 140065822

Anneka

imaged.com

naluwan

shutterstock · 47072353

kenny 1

shutterstock · 150360296

Lemuana

Studio One/ Shutterstock.com

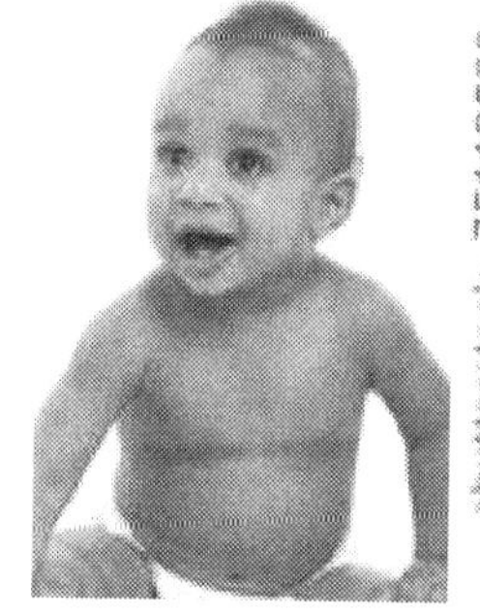

Samuel Borges Photography

Who among us does not marvel at the birth of a baby? We recognize this as a wondrous event, and it is.

A seven-pound (3.2 kg) newborn baby has been constructed over a nine-month period from two combined seeds, an egg and a sperm, plus quadrillions of the right atoms selected from the food the mother has eaten. It stands to reason that each of these right atoms has been found, counted, grasped, super-intelligently placed, and fastened to build each complex cell complete with its molecular machinery in operation. A baby is given trillions of newly constructed cells. Just **one** red blood cell, for example, is made of approximately 280,000,000 molecules of hemoglobin x 10,000 atoms per molecule = 2,800,000,000,000 (Twenty-eight hundred billion) right atoms per red blood cell. (See reference [1] Chapter 8).

The wrong numbers and/or the wrong atoms would simply not work. Cells have to be carefully and super-intelligently built. We in science do not have enough intelligence to build even one molecular machine for any cell, so why would we think that a theoretical process with no intelligence, could. I can't do this. Can you? Don't feel bad because even the best experts in the field cannot.

The volume, speed, precision, dexterity, and brilliance required to do this work is so awesome that it really should be personally appreciated, if not revered.

Since it happens so regularly and reliably, we can easily take it for granted. However, this IS an enormous amount of careful, super-intelligent, physical work that evolution is incapable of performing as it has no care or intelligence to use.

3. He Builds Our Astounding Brains

According to Professor Paul Reber of Northwestern University (and others) the human brain consists of about one billion neurons, each of which forms about 1000 connections and each neuron can connect with other neurons. As he stated in the June 2010 issue of *Scientific American,* this results in "...exponentially increasing the brain's memory storage capacity to something closer to around 2.5 petabytes (or a million gigabytes)."[7] Reber compares this to about three million hours of recorded TV shows. With this much storage, the recorder could play continuously for over 300 years.

Van Wedeen and L.L. Wald of the Martinos Center for Biomedical Imaging Human Connectome Project state that "The brain's many regions are connected by some *100,000 miles* of fibres called white matter, enough to circle the Earth four times" [8]

Jeff Lichtman, a neuroscientist and professor at Harvard, is studying brain compositions. He was interviewed by Carl Zimmer who then wrote in National Geographic, February, 2014, "So far the largest volume of a mouse's brain that Lichtman and his colleagues have managed to re-create is about the size of a grain of salt. Its data alone total a hundred terabytes, the amount of data in about 25,000 high-definition movies." They also indicate in the article that a mouse's brain contains about 70,000,000 neurons and a human brain contains about 1000 times that number. [9]

Douglas Axe, PhD, is an engineer-turned-molecular-biologist and director of the Biologic Institute. In his recent book, "*Undeniable, How Biology Confirms Our Intuition That Life Is Designed,*" he reminds us that, "*The human brain is different.... Being the most remarkable component of the human body, it is arguably the most outstanding physical invention ever to exist.*" [10]

This phenomenal computer, our brain, also had to be super-intelligently built of atoms from the soil, air, and water with God's phenomenal two-step process: Atoms in the soil built into food, then those same atoms in the food used to build our brain cells.

Where else would these atoms come from, and who else has the super-intelligence to assemble them in such a phenomenal manner?

4. He Builds Our Pets

Eric Isselee/Shutterstock.com

Susan Schmitz/Shutterstock.com

Using exactly the same mind-boggling, super-intelligent planning, vision, technology, precision, dexterity, and care, God builds all other creatures as He builds us. None of us can come anywhere close to doing this brilliant work.

5. He Builds Beautiful Flowers For Our Enjoyment

Have you ever marvelled at the beautiful colors, amazing fragrance, and functional design of some glorious flowers, and wondered how they could be made out of dirt? Does this not seem miraculous?

Flik47/shutterstock.com

6. He Builds Beautiful Fish

As God builds land creatures from the available atoms on land, so does He build beautiful sea creatures from atoms in the sea and in lakes and streams.

Vlad61/shutterstock.com

7. He Builds Beautiful Birds for Our Enjoyment

Do we marvel at the amazing colors of different birds, at some of their intriguing songs and calls, and at their wonderful ability to fly?

nattanan726/shutterstock.com

Stephen Russell Smith Photos/Shutterstock.com

Phillip Rubino/Shutterstock.com

8. He Works So Diligently for Us

Building each one of us plus all of our foods, pets, flowers, and so on, is no simple task, (the understatement of life).

Since the concept of "atomic biology" is relatively new, many of the exact details regarding the right numbers and types of the elemental atoms required to build each type of our various cell parts are yet to be discovered. However, the cell composition we *have* analyzed can give us some clues as to the work required in other cells.

In 2007 we calculated, with the help of C.J. Pallister, G.J. Tortora, and Max Perutz, that to manufacture the enormous number of replacement red blood cells for an average person, over 4900 quadrillion (4,900,000,000,000,000) *of the right elemental atoms every second* (see reference [1] Chapter 8) have to be found in our digestive system, picked up by our blood streams, delivered to blood cell construction sites, then sorted, selected, grasped, precisely placed, and fastened, with the special breath-of-life added.

Every second, each of these 2,000,000+ new red blood cells has to be manufactured and moved into our bloodstream with as many old ones being removed and delivered to our waste system and other areas.

This may be less than half the number of atoms that have to be precisely placed just for building our replacement red blood cells because, in the same second, a greater number of atoms (4900 quadrillion+ for fruit or vegetables including unused leaves, roots, etc.) have to be found in the soil in various gardens and fields, then sorted, selected, counted, grasped, then precisely placed and fastened to manufacture the food for each future second's red blood cells for each of us. This may bring the total to approximately 10,000 quadrillion of the right atoms that are brilliantly, faithfully, precisely, constantly, and carefully assembled for each one of us every second of every day, just for our replacement red-blood cells.

How can anyone believe that this enormous and super-intelligent physical work can be performed using no intelligence at all.

In addition there are at least 80 trillion (80,000,000,000,000) other cells in each adult that are being constantly and carefully sustained, maintained, repaired, or replaced.

There are so many other types of cells to build and sustain for each of us; – for eyelashes, hair, and skin cells of various colors, to heart, lung, liver, and brain cells, to name only a few.

God also makes herbs and other special plant items that have been used for healing by some groups for centuries. Many of our pharmaceuticals come from plants. There are even plant remedies which cure some types of cancer as noted by Suzanne Diamond, M.Sc. in an article titled *Canada's Amazing Anti-Cancer Tea!* [11] Another great article by Diamond outlining the special remedial effects of some of God's custom-built foods for us is *Humble Herbs Worth Their Weight in Gold,* published in a 2008 issue of Total Health Magazine. [12]

She has also written an informative book, *Nature's Best Heart Medicine,* that outlines new scientific discoveries revealing how flavonoids are beneficial for people with heart disease, high blood pressure, varicose veins, circulation problems, and more. [13]

We believe that determining the exact numbers and types of elemental atoms in each type of cell will be a great benefit in the development of more nutritional foods, and new pharmaceuticals to supply for use in the repair of cells that need to be healed. The information will also help in developing improved fertilizers that will contain all the necessary elements for optimal growth of our foods. There could be other attributes to the discovery of this information that would be beneficial for our health and healing.
The logic is simple and understandable when you consider the following aspects of Atomic Biology:

Every material thing, including each of our cell parts, is made of atoms. If we determine exactly how many atoms of each type of element are in each kind of our healthy cells, and we determine if any one of our cells is experiencing trouble because it lacks a portion of some elemental atoms, we could make sure there are enough of those types of atoms available to be placed, super-intelligently, into those cells for healing;

When we determine which of our foods contain an abundance of the types of atoms our unhealthy cells require, we can be sure to eat enough of those types of foods;

Conversely, if we have a tumor or other unwanted growth, we may determine the elements in the growth and find a way to prevent those elements from reaching the site;

When we determine what atoms make up each healthy fruit, vegetable, nut, etc., we can make sure that there is an abundant supply of those atoms in the soil where the atoms have to be selected to build the food item. Suitable fertilizer could be added if the soil is analyzed and found to be lacking enough of the right necessary atoms for the food items to be optimally built;

Using the right fertilizer for building good food cells is like using good foods to supply enough of the right atoms for our Creator to build good cells for our bodies;

This enormous, brilliant, reliable, careful, and trustworthy work of building our foods and our cells is performed for each of us free of charge every second of our whole lifetime.

These are just a few of the logical reasons for Thanksgiving. Many of us “give thanks” and say prayers at every mealtime.

In addition, even though we do not share it very well, God *does* manufacture enough food for all seven billion of us every day. Calculating the enormous amount of the brilliant work necessary to precisely place all the right numbers of the right atoms of about 60 different elements to construct this food, and then us from our food, could boggle one’s mind.

Of course, all of these atoms had to be created in the first place. And all these atoms contain electrons that must be moved around the nucleus at precisely the right speed virtually forever. This perpetual motion is not a natural phenomenon; perpetual, super-intelligent control of the energy being supplied is required.

Then, for us to exist, we need the following:

The right amount of sun to grow our foods and help us to see, to help keep us warm enough but not too hot, and to help change the tides to refresh our oceans;

Enough good air to breathe with enough oxygen atoms and other air-born atoms necessary for our life;

A system to take the carbon dioxide we exhale and convert it into oxygen for us to inhale which is exactly what our Creator helps plants and trees do for us;

In addition, are all the other essential details for life to be kept in balance and finely tuned (see Chapter 5).

It is easy to see that the vast amount of super-intelligent and caring work for us is absolutely awesome. In our upcoming textbooks, we will provide more details of this amazing work done continuously for each one of us, whether we deserve it or not.

Does God deserve our appreciation?

Chapter 2
Our Phenomenal DNA and RNA

What IS DNA ?

DNA, which stands for deoxyribonucleic acid, is crucial in helping cells perform their multitudes of amazing functions.

The main components of DNA include 4 bases:
Adenine - chemical formula $C_5 H_5 N_5$
Guanine - " " $C_5 H_5 N_5 O_1$
Cytosine - " " $C_4 H_5 N_3 O_1$
Thymine - " " $C_5 H_6 N_2 O_2$

Notice how similar the formulae are for these bases. As each one is being constructed for us using atoms from our digestive system via our bloodstreams, it is so critical that the builder be absolutely precise in the selection decisions, choices, counting, and placement of the right numbers of the right atoms, as well as fastening the right bases in the right sequences in programming our DNA for the various required functions in our cells. What super-intelligence, dexterity, and care this takes.

Some people credit chemical reactions for the assembly of cells and entities, but with a little thought we can tell that simply having C, H, N, and O hook up by chemical attraction would not make the super-intelligent choices and decisions for the prccise numbers of each elemental atom to choose and fasten per base. If the numbers of any of the atoms for any of these bases were not precisely correct, the whole DNA program would be corrupt causing a crash in the operation of the cell.

DNA is a super-sized strand of molecules (sometimes called a macro-molecule) reaching up to two meters in length according to Bruce Alberts and his co-authors in *Molecular Biology of the Cell, Fourth Edition.*[1] Almost all of our trillions of microscopic cells have an enormous DNA strand built into each one. The whole strand is wrapped into either the nucleus or the cytoplasm in each cell.

DNA macro-molecules are probably the most spectacular of all the machines constructed in our cells. Approximately 6,000,000,000 (six billion) base pairs have to be assembled within most of our approx. 100 trillion cells' DNA using the right numbers of the right types of atoms. The arrangement sequence is a huge super-intelligent code outlining our genetic information, heredity, a communication system, and instructions to help operate and maintain the cell. It is like an awesome, living, computer software program that has immense intelligent coding of our heredity from both parents (based on 23 chromosomes from our mother's egg and 23 chromosomes from our father's sperm). The programming also provides our individuality, and assists with the various specific functions for which each cell is designed, and wonderfully built of atoms.

Microsoft founder, Bill Gates, has stated: "Human DNA is like a computer program but far, far more advanced than any software we've ever created." [2]

Scientist I. L. Cohen said: *"At that moment, when the RNA/DNA system became understood, the debate between Evolutionists and Creationists should have come to a screeching halt."* [3] He is rightfully indicating that when the enormous complexity of RNA and DNA programming was discovered, the theory of evolution, which has no intelligence to program with, should have been dropped right at that time.

We know of no functional coding that was *made* without intelligent help, nor do we know of any intelligent coding that was *improved* without intelligent help. The tendency to deteriorate (the second law of thermodynamics) typically results when a coding change is not guided by intelligence. This applies to the immense coding in DNA as well.

Parts of our personal DNA strands (about 1%) are unique for each one of us. This feature helps authorities identify us if we are found somewhere without consciousness or identification, or if we are hurt, involved in a crime, being cleared for security purposes, and so on.

Other parts of these same DNA strands (roughly the other 99%) are virtually the same as in all our seven billion neighbors on our planet – for our eyesight cells, hearing cells, tasting cells, blood cells, liver cells, nerve cells, and so on, where the required functions are the same.

Then, of course, there are further differences for some of the cells for males and females. These differences are primarily for producing and raising children.

Each of our approximately 100 trillion cells is more complex than a whole city. There are several science projects listed online which display comparisons between functioning parts of a city to functioning parts of our cells. Rough analogies portrayed often consist of the following examples:

Cell Nucleus – City Hall which coordinates necessary functions;

Nuclear Membrane – City Hall fence with security guards;

Mitochondria – Power plants which generate controlled energy for cell operation and highly-regulated body temperature;

Cell Membrane – City border;

Cell Wall – a City wall with security guards;

Endoplasmic Reticulum – Streets and highways;

DNA – extensive detailed plans for building and running the city;

Nucleolus – Copy machine;

RNA – Plan copies for use around the city;

Chromosomes – Instructions from founders;

Ribosomes – Building materials transport;

Proteins – Workers;

Vacuole – Water tower;

Cytoplasm – Open space;

Protoplasm – Air;

Golgi – Post office;

Lysosomes – Recyclers and waste disposers.

All of these types of functions and more are necessary for running a cell. For a city there has to be a huge amount of intelligent planning, construction, servicing, maintenance, repair work, security, and so on. So it is with each of our cells.

While cities generally take decades to design and construct, our cells are super-intelligently built in a matter of seconds or minutes, from the time we started as an embryo.

The construction of our cities takes an enormous amount of human intelligence, knowledge, dexterity, skills, and physical work. The construction of our cells takes an even greater amount of super-intelligence, knowledge, dexterity, skills, physical work, plus lightning speed, and greater accuracy.

The amount of information programmed into the DNA in each of our cells is typically estimated as enough to fill more than a thousand sets of encyclopedias; or, if printed out in 12 point font, the printed line of information would reach from the North Pole to the Equator. These are rough examples commonly used by various scientists to give us the picture that the amount of intelligent coding in our DNA is gigantic.

The potential benefits of mimicking the applications of DNA technology, have become highly compelling to some scientists. For example, George Church and Sri Kosuri at Harvard University, and a group headed by Nick Goldman at the European Bioinformatics Institute, Cambridge, England have developed synthesized, non-living DNA for super-sized storage of digital information. The storage capacity-to-size ratio saves a dramatic amount of space, and is a safer way to ensure data accuracy in long-term storage; however, the process remains costly.

Our personal DNA is live, active, and functional. It is a life-giving gift, designed and built by our caring Creator, along with the rest of our body, mind, and spirit.

Here is an image of the shape of a segment of the strand of molecules that holds information within most of our cells. These graphic images of a DNA helix are expanded thousands of times larger than real life size.

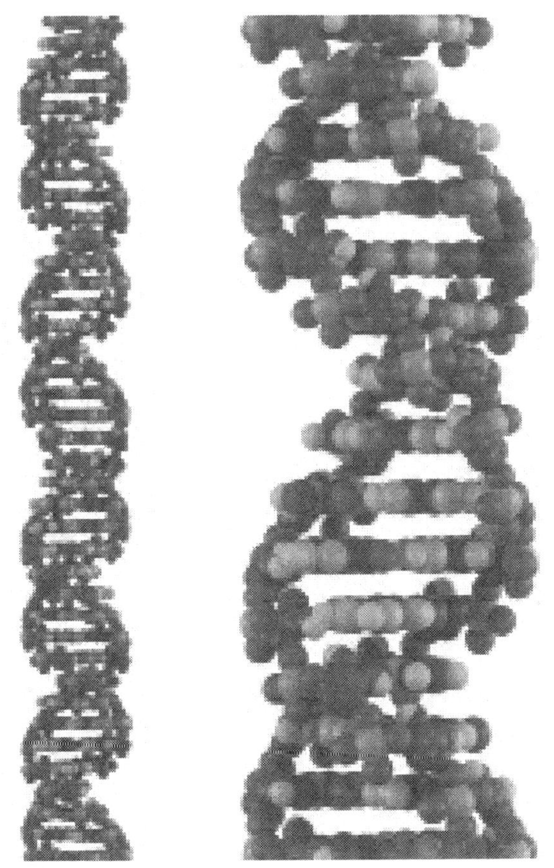

Mopic/Shutterstock.com

Our amazing DNA is designed with double helixes precisely built primarily of specially designed proteins. These proteins require so much intelligent programming at lightning speed, especially in our cells that have a life-span of only a few days, that this process goes far beyond the parameters of possibility for any unguided process.

It would be interesting to estimate the number of computer programmers it would take to program the functional coding for just one cell's DNA in four days for example. Of course, they would have to *know* the exact coding and how to make the right numbers of the right proteins out of the right numbers of the right types of atoms before beginning.

These factors point to the immense amount of brilliant, super-intelligent, physical work required to find, sort, select, grasp, precisely place and fasten all the right numbers of the right atoms, by a phenomenally intelligent, dexterous, and caring force. Of course, in the English language, we call this force "God".

Meanwhile, we just have to eat the types of foods that provide sufficient atoms of each required 59 or 60 elements for God to build all of our body's cell-parts.

The sooner we investigate the atomic needs of each type of our cells, plus the atomic content of fruits, vegetables, meats, and fish, the sooner we can match our recommended foods to our body's needs. Some food items might lack some of their normal elements if the soil they were grown in, lacked some of those elements.

Atomic analysis of field soil content, conducted to match the atomic requirements of our various food items, will reveal what fertilizer elements should be added to the soil for each crop.

On a topic related to our DNA, we must recognize that our cells do not just replicate themselves, as some people think. This can easily be proven by considering what would happen if the first cell in our development were a toe-nail cell. If cells only replicated themselves, then would we each be one big toe-nail? The answer is obvious.

The truth is that there are many different types of cells as we can see just by looking at ourselves. There are skin cells, bone cells, muscle cells, hair cells, eye cells, nose cells, ear cells, teeth cells, tongue cells, and so on. Each type has different designs, different elements, different structures, different DNA, different functions, etc.

More detail regarding the design, construction, and operation of our DNA will be developed for our upcoming college and university textbooks. (We seek science authors who may wish to participate).

With all of this remarkably detailed, wondrous work, carefully and awesomely built and programmed into our DNA, does God not deserve our appreciation?

What is RNA?

RNA stands for Ribonucleic acid. It is another macromolecule like DNA, that is essential for the functioning of cells in all known forms of life.

RNA is also composed of 4 bases, Adenine, Guanine, Cytosine, and the Thymine is replaced by Uracil - chemical formula $C_4 H_4 N_2 O_2$ (again having a very similar but slightly different chemical formula requiring super-intelligent and careful assembly).

Molecular biology tells us that the flow of genetic information is from our DNA through our RNA to the various "workhorses" or molecular machines in our cells. As mentioned above the information has to be super-intelligently programmed into our various cells' DNA in the first place.

Since different cells have different functions, so the DNA programming has to be different from cell to cell. Also, different workhorses have different jobs to do within each cell and the RNA molecules communicate the various work instructions to the various workhorses/molecular machines.

rRNA is a structural component of ribosomes which are molecular machines that help in the production of proteins within the cell. RNA molecules can also act as enzymes called ribozymes.

Super-Intelligent Decisions and Choices

This all leads to the following crucial question: for those who think that God designed and built the amazing biological systems for humans into Adam and Eve and has never intervened in our construction, sustenance, maintenance, or repair since, how do we account for all the super-intelligent decisions, choices, and physical works necessary to switch from assembling a blood vessel's outer membrane cell to a muscle cell or a nerve cell or an eye part cell right beside it, for example?

In the human body, there are about 60 different elemental atom types used to make all the various parts for all our various cells.

There are virtually countless decisions and choices to make just in selecting the right atoms by the right numbers to make all the various molecules for all the various parts of so many different cells.

Let's go back to assembling and programming just the DNA macro-molecule in just one of our skin cells for example, knowing that this contains the equivalent amount of information to roughly 1000 sets of encyclopedias. Think of just the number of choices and decisions necessary for selecting and precisely assembling the right numbers of the right atoms from our blood stream for that one part of that one cell. Then making the decisions and choices necessary for assembling the many other different parts for that one skin cell; then switching to building a muscle cell right next to the skin cell with a totally different function.

It is really all these super-intelligent decisions, choices, and physical works that rule out everything but a super-intelligent force for creating living cells and entities like us.

We and our governments in these four nations, call this brilliant and caring force, our "Creator God". "Thanksgiving

Day" is a national day of appreciation for His phenomenal work, sacrifice, provision, and care for us.

Chapter 3
Our Amazing Systems and Senses

As we have established from about grade five science, every material thing is made of atoms. This includes all parts of our body. Where do the atoms for building us come from? From the food we put in our mouth and air that we breathe in. In our beginning, the atoms came to us through our umbilical cord attached to our mother's womb and blood system.

And where do those atoms for our food and oxygen come from? From the soil of gardens, fields, and orchards, and the fresh air from plants that take in carbon dioxide and produce fresh oxygen. Then what has to happen after we have eaten and breathed-in these atoms from our food and air?

What else but the re-sorting of the atoms in our digestive system and lungs to find the right numbers of the right atoms for building our various cells, grasping them, precisely placing them, fastening them in their proper place, programming the DNA and RNA, breathing new life into the inanimate atoms, and placing these new cells in the proper place in our tissues, organs, bloodstream, nerves, replacement cells, etc.

All of this activity takes a phenomenal amount of super-intelligent inventing, designing, planning, and physical work – 24/7 – for our whole life.

If that is not enormous loving care for us, what other reason would our Creator have for doing all this work for us?

People who have figured out that God loves them, even if they did not know these details, receive a special benefit through their relationship with the Creator that cares so much for us all. Showing their personal appreciation for God's work and care for all of us might be a form of religion, but the rest of this knowledge is pure science.

The detailed construction and functions of *each one* of our systems and senses could fill a large book, so we are just going to provide a brief overview of each marvelous one.

OUR SYSTEMS

1. Our Digestion System

After we eat our food and breathe-in air, the huge task of sorting the right atoms from the unneeded ones begins in our digestive system and lungs.

The work of breaking our food down into smaller and smaller pieces begins in our mouth. This is why chewing our food properly is so important. This also helps to mix the digestive juices in our mouth with our food to begin the digestion process.

The next step of further breaking down our food into smaller pieces is done in our stomach where stronger digestive juices are added.

Remember: God has to pick the right numbers of the right atoms out of our food so that He can begin building new cells for us to grow up when we are young and to replace worn-out or damaged cells as needed at any age. He also has to deliver certain atoms from our food out to all our cells for their energy supply, sustenance, maintenance, and repair.

Some of the cells need repairing because they were damaged by cuts or scrapes or some other abuse. Many need replacing, like our blood cells, after about 120 days of absorbing and delivering oxygen and other nutrients to all the other cells around our entire body, and removing the carbon dioxide from our lungs.. For an average-size adult male (based on a 70 kilogram or 154 pound male) this requires about 2,300,000 new red blood cells to be built and delivered into our bloodstream *every second* of every day.

Each red blood cell is made of about 280,000,000 molecules of hemoglobin and each of these requires about 10,000 right atoms. Doing the math, approximately 2,300,000 x 280,000,000 x 10,000 = 6,400,000,000,000,000 (6,400 quadrillion) right atoms every second of our adult male life have to be found in our food, sorted, selected, grasped, precisely placed, fastened, DNA programmed, with life breathed into the inanimate atoms, the new cells delivered into our blood-stream, and the old ones removed and placed into our waste system (see reference note [1] Chapter 8).

Building the foods to supply all these atoms requires the handling of more atoms than that huge number, for making the roots, leaves, and skins which are thrown away. Therefore, this may bring the total to more than 12,800 quadrillion atoms per second that must be carefully handled just for our replacement red blood cells, although there may be a percentage of atoms recycled that has yet to be determined.

We have another roughly 80,000,000,000,000 (80 trillion) cells that also have to be maintained, repaired, and replaced to keep us alive.

The numbers can be a bit mind-boggling, but all this is said to indicate how much each one of us is continuously cared for by our Creator.

Of course, *we also* have to take some responsibility for our own health and well-being. We are given the freedom to choose how well we will do this. What we put into our mouth or breathe-in is important because this is the source of the atoms we need for our best health. We need healthful foods and beverages in about the right amounts – -not too much or too little. We do not need "poisons" or other substances that will abuse the supply of atoms that God needs to work with for our life and benefit.

We shouldn't expect to stay healthy if all we consume is coffee, cigarettes, donuts, booze, drugs, polluted air, and the like.

Reasonable exercise helps our digestive process as well as our muscle capacity, circulation, energy supply, attitude, and overall wellness. Our common sense and self-care play an effective roll in our quality and enjoyment of life. Have lots of physical fun.

2. Our Respiratory System

The quality of what we breathe-in is also important to our health because some of the atoms God needs to use for building, sustaining, maintaining, and repairing our body come from the air we breathe.

Unfortunately, many of the air-pollutants in our cities adversely affect the quality of this source of atoms for His works for us. And inhaling cigarette smoke and poisons like cocaine, and other drugs, will ruin our health over time. This, of course, brings suffering of one kind or another.

As we are told repeatedly, exercise and fresh air are important for our health. The deep breathing brought on by exercising increases our crucial oxygen supply and helps eliminate the toxins we have breathed-in.

Part of God's great design and physical work for our lives is to use the carbon-dioxide in the air we breathe-out for the construction and use of trees and plants. These, in turn, give out oxygen which God then uses for our construction, sustenance, maintenance, and repair.

It is important to understand that our governments cannot solve all of our problems, and God won't either. They should not be expected to do so. We are blessed with many free choices but we need to use wisdom in choosing the way we manage our lives; otherwise we will have unnecessary troubles. Our choices bring consequences, good or bad, for us.

It is up to us to help take care of ourselves, our nation, and our allies.

3. Our Blood System

As mentioned above, God has to perform an enormous amount of super-intelligent work for us just to replace our worn-out red blood cells. Sorting, selecting, grasping, precisely placing and fastening more than 4900 quadrillion atoms per second just to build our replacement red blood cells, goes far beyond the parameters of possibility for any unguided process like evolution. Consider the point that humans, the most intelligent beings on Earth, cannot build even *one* live molecular cell machine from chemical elements given any amount of time.

Our blood system has many crucial jobs to perform including, but not limited to:

1. Collecting oxygen from the air we breathe into our lungs and transporting it to every cell;

2. Delivering the carbon dioxide our cells must exhaust as waste back to our lungs to be exhaled;

3. Gathering many kinds of atoms from our digestive systems and delivering them as nutrients to each of our cells for God to place for their energy, nourishment, maintenance, and repair;

4. Delivering dead or worn-out cells to our waste system;

5. Helping to regulate our body temperature which is crucial to maintain within a very small variation in every human on Planet Earth;

6. Delivering God-designed and built healing fluids to our wounded areas;

7. Helping our immune system to fight invasive bugs and viruses;

8. Disposing of these foreign invaders to our waste system;

9. Delivering of special atoms from pharmaceuticals taken to places within us where they are required to help fight colds, headaches, and other illnesses;

10. Blood-clotting to prevent hemorrhages;

11+. And more.

4. Our Waste System

After our digestive system has broken down our foods into small parts, and God has sorted, selected, grasped, and taken all of the right atoms He needs to work with for our cell sustenance, maintenance, repair, and replacement, He leaves the unneeded atoms in our digestive track (intestines) as waste for elimination.

Disposing of our waste is another crucial part of our life systems. Additional waste includes worn-out cells, destroyed sickness 'bugs', liquid wastes, toxins, and so on. This is another marvelous part of the design and operation of our body.

Regular elimination of this toxic waste is very important for our health.

5. Our Nerve System

This is another phenomenal part that God builds into us to aid our body functions.

Nerves are the amazing hard-wiring for all the electrical impulse messaging necessary for our brain functions, muscle functions, eyesight, hearing, touching, tasting, smelling, and operating of our conscious and sub-conscious activities.

According to scientists Van Wedeen and L.L.Wald, each human brain contains approximately *100,000 miles (161,000 kilometres*) of nerve fibres. (See ref. note [8] Chapter 1).

Using separate nerve-fibre lines, God also connects all of the following to our brain:

- every one of our muscles;
- all parts of our speech system;
- every skin-cell sensor for our sense of touch and pain-warning;

- all parts of our other senses, including our eyesight, taste, smell, and hearing;
- all our organs for their functioning;
- and, of course, the fantastic communications and function systems within our brain itself.

God also builds more spectacular messaging systems into each of our 100 trillion cells to help guide all of the amazing work that goes on in each one. Just reading and responding to the function instructions within the DNA to guide the work in each cell, boggles one's mind.

There is such amazing complexity in each little part of our body that the super-intelligent brilliance essential for its design and construction is virtually overwhelming to this humble scribe. How about you?

6. Our Immune System

Here is another amazing system that God has designed and built into each of us. Because we are not able to keep perfectly clean, and as there are both helpful and harmful bacteria all around us, He gives us an internal system to combat the bad bacteria, germs, and viruses.

We have learned that we can strengthen this system against some germs by using vaccinations.

Actually, not trying to stay perfectly clean or sterile is helpful because exposing ourselves to a certain amount of germs acts, to a degree, like a vaccination, as long as we do not get too much.

The other help we can provide is to make sure that we eat healthful foods and beverages, get fresh air, and exercise. This takes a little planning and will-power but these are probably the best investments we will ever make for our body.

7. Our Reproduction System

This is the most phenomenal and spectacular system of all because this is how our Creator builds all our magnificent parts, systems, and senses, then connects them all with blood vessels and nerves and covers them with skin to make this fantastic, living machine in which we live.

The marvelous work of building each one of us started when our mother was born. At that time God had constructed about 400,000 egg cells into her little ovaries for future use.[1] Although she lost one or more eggs with each menstrual cycle, one of the remainders was available for fertilization and the beginning of our life when the right time came for our construction to begin.

These egg cells were constantly sustained with nourishment and maintained in good condition so that one of them could provide its special role in our assembly. The job of that one special egg cell was to provide about half the design plans for the construction of all our parts, and unique features, looks, and traits when our time came to be created.

When the right time came for our mother to be able to bear children, God would send one or more of these egg cells from their storage area in her ovaries, down towards her baby construction site – her womb – in preparation for possible fertilization which would begin the baby construction process.

Normally, one egg cell would be released from our mother's ovaries every month. If no fertilization took place within a few days, the egg cell would be flushed out of her womb along with some special blood that was placed there to help with the possible baby construction process.

This monthly process continued until it was our time to be created. Until then, our mother might have made some arrangements to avoid becoming pregnant by waiting patiently for the right time or by using some kind of birth-control pill or device. But our month to be conceived finally came. Then it was time for our father to get involved.

He had to provide a fertilizer-cell, called a sperm cell, to unite with our mother's egg cell of the month. This was the easy part for him; then he was supposed to, and hopefully did, support the family members he helped to produce.

Once the mother's egg cell and father's sperm cell united, our Creator started the super-intelligent task of building us from the "dust" with His phenomenal two-step process. He built-in our unique features in accordance with the plans that were built into the original egg cell and sperm cell.

All the food atoms that God did not need for building us, and for sustaining, maintaining, repairing or replacing our mother's cells, were disposed of through our mother's waste system.

Building a human baby generally takes God about nine months. There is currently a great video called "Conception to Birth – Visualized" from medical-image-maker Alexander Tsiaras at a TED Talks INK conference. Hopefully it is still showing as you read this passage. The online address is:
www.ted.com/talks/alexander_tsiaris_conception_to_
birth_visualized , or you may be able to watch it on YouTube.

The phenomenal sequence of construction from the fertilization of an egg to the birth of baby goes about like this:

- Day 1: The mother's egg is released from one of her two ovaries and is captured by one of her two oviducts.
- Shortly after the egg enters the oviduct, it may be met with some of the sperm. The egg is designed with complex

sensors and mechanisms to allow generally only one sperm, and nothing other than human sperm, to enter the egg.

- The fertilized egg is moved down the oviduct and begins to divide internally as it is moved along. After the first division, if the two cells separate and live, identical twins will begin to be constructed. If two eggs happen to be fertilized by separate sperm, non-identical twins will begin to be constructed. For our purpose, we will stay with the construction of a single child.
- Day 6: The fertilized egg cell is fastened to the wall of our mother's womb with filtered access to nutrient atoms from our mother's blood vessels (more brilliant work).
- Day 25: An amazing amount of physical work has been performed, including the gathering of the right numbers of the right atoms from the nutrients delivered by our mother's blood stream from the digesting food in her intestines to the uterus (womb), then the selection, grasping, precise placement, and fastening of the right numbers of the right atoms to build the embryo to the point of having a recognizable head, spine, and heart.
- Day 56: The super-intelligent force we call God has already assembled all of the organs to an early stage.

From there on He has to complete the construction of the baby which involves the precise assembly of approximately 4.2 trillion cells. He makes collagen tissue, which is stronger than steel by relative weight, to hold the fetus and its marvellous parts together and in their proper place.

All of the cells under construction have their own DNA instructions built into them and programmed with phenomenal amounts of information (see Chapter 2 for more about our DNA).

Of course, all the atoms used for building the cells have no life of their own, therefore, God's crucial breath-of-life must be added to make each cell live and function.

What parts have to be built and hooked up together to make a baby?

There would need to be some skin to hold everything together, wouldn't there? Incidentally, our skin is the largest organ of our body.

God had to build a brain and heart for us early on in our construction; blood and blood vessels; liver; kidneys; stomach; intestines; pancreas; gall bladder; eyes; ears; nose; mouth; hair; arms; hands; legs; feet; fingernails; toenails; skeleton; muscles; lungs; nerves; wiring; and so on. Plus, He had to program our DNA in almost all of our cells so that everything functions properly together. He also had to make several organs that are different for girls and boys.

What an enormous amount of beautiful, phenomenal, super-intelligent effort He put into designing and building each one of us. What wonderful care He provided and continues to provide to us in making our foods and reconstructing our food atoms into our new cells for our growth, maintenance, and repair every second of every day.

An average-sized human baby human weighs about 7 pounds (3.2 kg) at birth, and then over the next twenty years, grows to be an adult. Adults are built into various shapes, sizes, and colors, as we know, but an average male in the four nations is about 154 pounds (70 kg). This is about 22 times the size of the little baby at birth.

To build an adult of this size, God has to construct about 100 trillion (100,000,000,000,000) cells compared to the approximately 5 trillion cells (5,000,000,000,000) cells when we were born. As He builds us, of course, He has to provide His breath-of-life to each cell as He completes it.

It took Him about nine months working super-intelligently, quickly, and brilliantly – twenty-four hours a day – to build each one of us to the point of birth. He initially got the atoms from our mother's womb until He had enough to build an umbilical cord, which He then attached from our little body to the special blood vessels in our mother's womb.

How awesome is that?

Incidentally, as God has to begin building each baby with a sperm and an egg, it would not have been difficult for Him to build both of these in the Virgin Mary's womb. (Just one of my wandering thoughts).

OUR MARVELOUS SENSES

1. Our Sense of Sight

What a fantastic invention this is, and what a phenomenal, super-intelligent construction project.

Our eyes, the most valuable and intricate of all our sense organs, provide vision for us. Looking about you, you can see all that is around you. Can you imagine what it would be like to be blind? Blind people adjust to life and heighten other senses including hearing and touch, but they miss so much that we with vision take for granted. Being able to see the world around us is of great value, helping to make our life interesting, safe, and enjoyable.

These small round organs have many parts, all of which are responsible for sight, protection, maintenance, and clarity. People are sometimes born with different-shaped eyes that can distort vision but this can now be corrected by man-made lenses or surgeries.

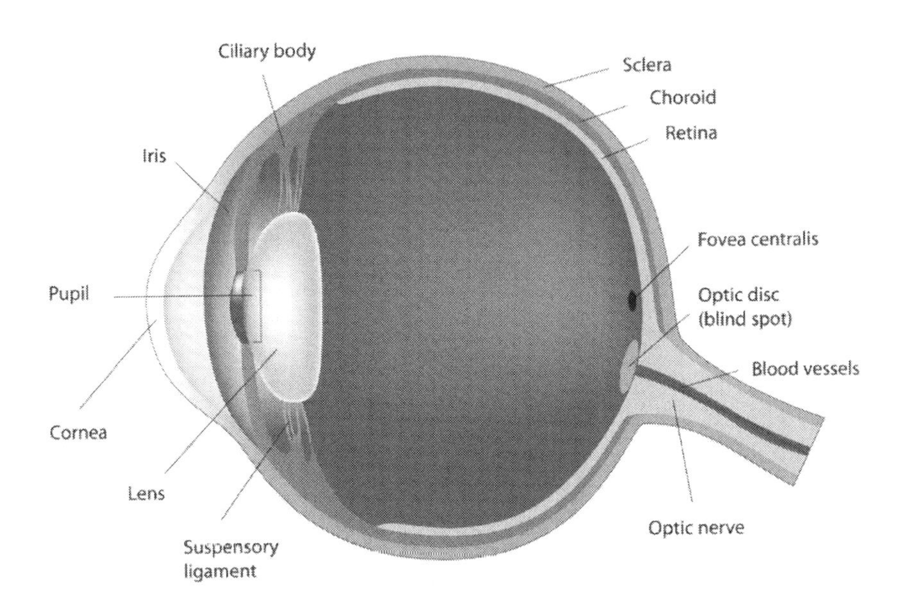

Alila Medical Media/Shutterstock.com

Protective structures for the eye include the orbital bone which surrounds the eye; eyelids; eyelashes; eyebrows; and the tears that come from the lacrimal glands with ducts that keep the eyes lubricated, moist, and clean. There are even sebaceous glands which produce oil to keep the eyelids from sticking together.

The eyeball is fitted with muscles to support it and give it movement, a vascular segment to give it circulation and the nerve portion called the optic nerve, located at the back of the eye. Any damage to the optic nerve can cause partial or total blindness.

There are numerous amazing parts to the eye when examining it from the outside to inside. The first layer, sclera, is the white part of the eye that gives it shape, stability, and protection. The optic nerve passes through the sclera at the back of the eye. The cornea is a transparent coat that covers the colored part of the eye and the pupil.

The second layer contains muscles, blood vessels, and the iris. The pupil is the black hole in the middle of the iris where light enters the eyeball. The iris regulates the amount of light entering the eye by dilating or contracting the pupil to adjust for the brightness or darkness of the viewing area. Another marvelous invention, design, and construction project here is the system for altering the blood to make it clear in this area so that the blood cells do not interfere with the light coming in through the pupil. Even the structure of the tissue here is changed from a rope type of assembly to an open weave. This also improves clarity for the light entering our eyes. How brilliant is that?

The third layer of the eye on the inside is the retina whose primary function is to register images. God has built multiple neurons into our retinas which may be referred to as rods and cones. They can register all shades of color and images and then accurately emit the necessary electrical impulses to our brain for translation into understandable images.

Another wonderful invention built into our sight system is the coordinated double image from our two separated eyes to give us depth perception, distance, and a more interesting three – dimensional imagery.

The interior of the eye has an anterior and posterior chamber containing fluids and gels between which sits the transparent body called the lens. For vision to occur, light must pass through all of these structures. The lens and pupil accommodate refraction of the light waves to converge on the retina with the amount of light that allows us to see.

These are just a few of the basic principles about sight; hence, you can well imagine the complexity and divine hand that goes into the construction of our eyes for our marvelous vision. This is something that we can truly appreciate in our beautiful world.

2. Our Sense of Hearing

The ear is another organ that is amazingly created. Since we usually see only the outer ear, we might not consider how the ear functions with all its intricacies involved. Most people do not realize that the ear has two important functions: to register and transmit auditory sensations to our brain (hearing), and to maintain our equilibrium (keep balance), both of which are very important to our lifestyle. Fortunately, a variety of devices are available to assist hearing. When something goes wrong with our equilibrium, we are often not so fortunate. Our ears are designed and built to contain receptors that register sound waves and accurately transmit the information to our brains. Our ears also have built-in receptors which help us keep our balance.

The ear consists of three sections: the outer (external), the middle, and inner ear. All parts are vital and can be extremely disabling if damaged in any way.

The visible part of the ear is called the pinna. Attached to the head by muscles and ligaments, it is shaped specifically to pick up sounds and protect the inner working mechanisms of the ear. From the pinna, an external canal leads into the eardrum. The highly sensitive skin inside the canal contains many mini-hairs and cerumen (ear wax) to prevent foreign bodies from entering the internal ear. At the end of this canal lies the eardrum or tympanic membrane, also covered with skin with a mucous membrane inside. Since this membrane is easily penetrated, great care must be taken not to push anything down the ear canal.

The middle ear is an air-filled cavity that communicates through the mastoid cells in the temporal bone and then into the auditory (Eustachian) tube leading to the throat. This serves the purpose of equalizing pressure on both sides of the eardrum to prevent rupture. Three very tiny bones called the malleolus, incus, and stapes cross the middle ear connected by synovial (hinge) joints. They connect the eardrum to the inner ear and are attached to the bone by ligaments.

The inner ear, called "the labyrinth," consists of cavities called vestibules, a cochlea (coiled hollow tube), and semi-circular canals. Fluid surrounds this area while all its parts are connected by many sacs and canals. Tiny hairs, both inside and outside the cochlea, help transmit sound.

In basic ear physiology, sound waves are caught and directed by the pinna, then pass down the ear canal to the eardrum. Low-frequency sounds cause slow vibration, while high-frequency ones cause rapid vibration. The three little bones vibrate accordingly and pass the sound waves through the inner ear system to the cochlea. For hearing to result, inner ear mechanisms, pushing back and forth, stimulate the auditory nerve to carry sound signals to the correct part of our brain.

Equilibrium is controlled by three semicircular canals located in the inner ear. They face in different directions and contain many hairs of different lengths and consistencies. A gelatin-like fluid flows through and moves the hairs to stimulate sensory neurons which activate nerves to help maintain balance if your head position changes.

With so much intricacy and functional interdependence, do you agree that it would take a super-intelligent power to invent, design, and construct this marvelous hearing and balancing mechanism for each one of us?

3. Our Sense of Touch

Our brain can detect the exact point of touch on our skin. Our sense of touch works when our skin comes in contact with different objects that register hot or cold, stillness, vibration, texture, tickling, tingling, pain, pressure, and so on.

Our skin, the largest sense organ in our body, is composed of three layers. The epidermis is the outside layer, which we can see. It contains many sensitive cells that receive information about the environment that it touches. The next layer is the dermis which contains touch receptors that have more neurons in some areas than others, and hair follicles and nerve endings. More epidermal cells are formed here and rise to the top to replace any dead epidermal cells. Oil and sweat glands are also found in the dermis layer. The third layer is a fatty one which helps keep heat in our body and helps prevent damage to underlying structures. Intact skin also helps prevent germs from entering our body.

The sense of touch is a great benefit to us. By touching items we can determine size, shape, temperature, texture, and many other attributes. The sense of touch also relates us to our surrounding world, such as how we walk, sit, balance ourselves, etc. For our safety we need many nerve endings and receptors to recognize hot, cold, and pain. This helps us to avoid burning, freezing, or otherwise damaging our flesh. Detecting vibrations and even a loving touch like a hug or a kiss are helpful and enjoyable. The desire for a touch of affection is very strong.

The brain can detect the exact point of touch. Our fingertips are particularly sensitive. When they are moved over an area, they can give exact details that no other skin area can. Pressure sensations on deeper tissues last longer than a light touch.

Pain sensations and their receptors are found all over our body. When hot, cold, pressure, or vibrations reach a certain level of intensity, pain can occur. Some of these may be caused by burns, cuts, insect stings, or bites. Excessive stimulation of any of the touch areas can cause pain which can help minimize damage.

It is evident that our skin senses, brilliantly designed and built, are all carefully and precisely "wired" into our brain from every tiny area. They are an amazingly beneficial part of our body.

4. Our Sense of Smell

It is easy to think that smell just tells us of a change in our environment, either good or bad. However, many critical things are identified by smell, such as smoke from a house fire or food burning on our stove. Therefore, it can be debilitating or life-threatening to lose that sense.

All of our sensory organs have receptors necessary to function in their own special way. The receptors for the nasal cavity are on either side of the septum, the dividing centre of the nose. Epithelial cells – which produce mucous – keep the nose, olfactory glands, and the air going to our lungs, importantly moist. The olfactory glands contain cells with bipolar neurons. The ends of these cells contain multiple hairs that react to odours and stimulate the olfactory pathway. This pathway leads to the cerebral cortex in our brain where it is determined that odours are present. Thus, we smell. This whole process happens in about two seconds.

Before we can smell a substance, it must enter into a gaseous, water-soluble state so that it can enter the nostrils and dissolve. Taking a deep breath and sniffing increases the smell. The membranes of the olfactory hairs are mainly lipid (water-insoluble). To initiate the impulse to smell, the substance to be smelled must penetrate that lipid layer. Many different theories try to explain the complexity of how the nose smells what it does.

Any problems along this information route can cause disturbances with our sense of smell. Loss of smell can be a truly distressing experience.

As with our other senses, you can well imagine the divine construction required to create a remarkable function like smelling.

5. Our Sense of Taste

The receptors for our taste – also called gustatory sensations – lie in our taste buds. Each taste bud contains multiple taste cells that allow us to recognize the flavours of sweet, salt, sour, bitter, and savoury. Each taste bud, of course, is "wired" to register and transmit the information to the correct part of our brain.

The taste buds are located under the mucous membrane of the tongue in the rough, wart-like surfaces called papillae. Of various shapes and sizes, these papillae are able, by their location, to identify the five tastes. The large papillae at the base of the tongue contain thousands of taste buds. On the tongue's upper surface are thousands of other smaller papillae. Each taste bud identifies different levels of intensity. As with other body cells, the taste buds are replaced once a week. These papillae contain a small opening connected to a saliva gland which provides this initial digestive solution to what we eat. Saliva also helps sensitize our taste buds, thereby increasing our ability to appreciate the taste of our food. From there, nerve cells identify the substances we have ingested. After contact with the food eaten, the taste receptors react rapidly. Smell is brought into play to help the taste buds. The combination of odours and taste involves a psychological adaption in the brain so that clear taste can occur rapidly.

The front of the tongue recognizes sweeter tastes, the sides of the tongue detect salt, savoury, and sour flavours, and the back picks up bitter tastes. Since all people are different, the intensity and pleasure or displeasure of each taste is individualistic.

In looking at all five of our senses, isn't it truly amazing how marvelously our Creator has made each one of us?

Chapter 4
Our Incredible Molecular Machines

What are the Molecular Machines within our cells?

The more we learn about the amazing variety of phenomenal works carried on within our cells, the more we realize there is no way cells can be built without using super-intelligent capabilities. The James Tour Group at Rice University has developed a number of relatively simple molecular machines, as have the 2016 Nobel Prize for Chemistry winners, Sauvage, Stoddart, and Feringa, as well as other groups working on similar projects.

"The first step towards a molecular machine was taken by Jean-Pierre Sauvage in 1983, when he succeeded in linking two ring-shaped molecules together to form a chain, called a catenane. .... The second step was taken by Fraser Stoddart in 1991, when he developed a rotaxane. He threaded a molecular ring onto a thin molecular axle and demonstrated that the ring was able to move along the axle. .... Bernard Feringa was the first person to develop a molecular motor; in 1999 he got a molecular rotor blade to spin continually in the same direction." [1]

For the James Tour Group, it took over a decade of expensive research and development to produce their first simplistic nano-machine made of just one molecule, under unnatural conditions. It can actually move with the help of stimuli such as ultraviolet light impulses. The assembly and testing of some of these sub-microscopic units has to be performed on a flat gold surface heated to between 170C and 225C. [2]

Fortunately, the many, almost infinitely more complex molecular machines constructed for us within each of our cells, are created at body temperature without the need for ultraviolet light or flat gold surfaces, specialized equipment, high-powered microscopes, etc. plus unacceptable lengths of time.

Obviously, there is far more intelligence involved in our cell construction than we in science can come anywhere close to at this time, as we can only build extremely simple units by comparison to our Creator's handiwork. We also cannot build any significant single part of a cell from raw elements, and we cannot provide the "breath of life" to anything.

This is why we have included in Chapter 5, one of the principles of life under "Dead dogs don't bark." Although the dog has every needed atom and molecule precisely assembled and placed for its eyes, ears, nose, teeth, heart, legs, brain, etc., without the "breath of life," it is not going to move one millimeter.

So why would anyone think that something with no intelligence at all, like evolution, could do all this super-intelligent life-causing work?

The theory of evolution as the cause of life, should now be considered ***"falsified."***

It is past time for science to "go where the evidence leads", especially in this field of constructing living cells and entities.

Many of the molecular machines in our cells, are irreducibly complex. This is another strike against the theory of evolution as the cause of life.

What this means is that these molecular machines could not even start to function over a long period of progressive development without having all their parts in place from their beginning.

Let's consider Darwin's theory of Origins whereby there was (theoretically) a common ancestor for all species at the base of his "tree of life". The theory surmises that an original cell was spontaneously created in a primordial pond. This cell would have needed the ability to find and process nourishment for its survival, to reproduce itself, to pass on its best traits and make many improvements to its design over many generations, eventually becoming fish, birds, plants, animals and people. Darwin was unsure how this could happen but assumed that it did happen and produced his theory from there.

There were many things that were not known about atoms, molecules, cells, and life at that time, 1859, and he admitted that there were a number of ways his theory could "absolutely break down."

According to Dr. Stephen Meyer of the Discovery Institute, the *simplest* known cell today needs 482 proteins and 562,000 bases for its DNA. (See ref. note [2] Chapter 5).

After long study, many remaining evolutionists, as well as former evolutionists agree that an original reproducing cell just happening by chance circumstances, is virtually impossible.
Now, we are learning more about the phenomenal complexities essential for the many different molecular machines assembled within each of our cells to help them function.

Of course, all of these various phenomenal molecular machines have to be super-intelligently constructed within each of our cells, using the right numbers of the right types of atoms, all precisely assembled and fastened in proper sequence, and with the "breath-of-life" added to make them functional.

The communication system within each cell has to be super-intelligently constructed and guided as there are constant needs for choices and decisions to be made to help with each cell's proper functioning. For example, choices and decisions have to be made as to what nutrients are needed to be extracted from the connected blood vessels, where these nutrients need to be placed, delivering them there, what parts of the cell need repairing or replacing and when, how to do that particular bit of work, when the entire cell needs replacing, etc., etc., etc.

We won't go into a description of all the types of molecular machines and their sub-varieties in our cells, but a few examples are:

Kinesin is a marvelous little workhorse made for us within most of our cells, to transport various cargo loads from one area to another where and when needed. It literally has two 'feet' to walk along a strand concurrently built ahead of it from where it has to fetch a particular cargo to where this cargo has to be delivered. It has two little 'hands' to hold the cargo as it flips one foot in front of the other along each newly built strand. There are animated videos of these amazing live machines online. If still running, see on YouTube, *The Workhorse of the Cell: Kinesin* by Discovery Science News.

Myosin can also deliver cargo in our cells, and a variant of myosin helps our muscles to contract by converting chemical energy to physical energy for that purpose.

Proteasome is a large group of proteins that act as waste disposer machines. When cell proteins are damaged or worn out, they are tagged, unfolded, and cut into small bits called peptides before the proteasome disposes of them through the cell portal and into the blood stream to be delivered to our waste system.

Lysosome is similar to a proteasome, but it breaks up a whole cell that needs to be delivered to our waste system.

Calcium pump is the molecular machine that pumps calcium ions across our cell membranes using a four-step process.

Ribosomes work with RNA in our cells to help in the manufacturing of proteins for our DNA, etc.

Mitochondrion contains many enzymes which help to convert some of our food nutrients delivered to our cells, into usable energy.

ATP Synthase is a complex molecular machine with a mass equivalent to that of about 500,000 hydrogen atoms. It is super-intelligently constructed to operate at about 6000 rpm within the membrane of mitochondria. Its purpose is to generate energy for physiological reactions like muscle contraction, for example.

As mentioned, there are several different molecular machines super-intelligently constructed, sustained, maintained, repaired, and replaced within most of our cells.

Casey Luskin of the Discovery Institute has compiled a list of 40 of the different types of molecular machines constructed for our life. [3]

The more we learn about the Life given to us, the more fascinating and gratifying is the wondrous care provided for us.

We just have to select good foods and health habits so that our Creator can help us have our best well-being.

Chapter 5

Moving Darwinism, Neo-Darwinism, and Macroevolution to the History Department

With input from 20 PhDs, 3 MDs, 9 DScs, 3 Mathematicians, 2 MScs, and 8 Independent Researchers, over a 30 year period, we believe we have made a strong case for falsifying Evolution as the cause of life. We have also made the case for the essentiality of super-intelligence and super-dexterity to build every living cell and entity.

There are lists of thousands more scientists who have openly declared their skepticism of Evolution as the cause of life, and God only knows how many other scientists, teachers, professors, and medical professionals who dare not make their skepticism known for fear of losing their jobs.

To provide clarity, we need specific definitions of the terms "Darwinism", "Neo-Darwinism", and "macroevolution" that we are dealing with in this chapter. Various dictionaries and universities provide different definitions but they all contain the similar basic aspects. These terms are not to be confused with "micro-evolution" which is simply a change in details like hair color, skin color, eye color, size, and minor mutations from generation to generation, but with no change of "kind" of creature or entity. There are definite boundaries to the types of changes our bodies allow through micro-evolution. We will also show that no cells are constructed without super-intelligent work, and this includes mutation cells.

The "evolution" we are referring to is the kind taught in our government-funded schools, colleges, and universities. The three forms of evolution, "Darwinism", "Neo-Darwinism", and "macroevolution" – are all similar theoretical explanations of the origin and cause of life and different species whereby all species and kinds originated and descended from a common ancestor. The theory of evolution surmises that this first organism was formed by an unguided, random chance happening billions of years ago. "Natural selection" is included as an aspect of evolution as an extension of the concept of "survival-of-the-fittest" plants and creatures. Theoretically, it accounts for the survivable mutations from the first organism into all the other kinds of plants and creatures. These changes happened, theoretically, through random mutations that provided changes and improvements in ascending entities. There is no guidance, purpose, design, or intelligence involved with this theoretical process.

If this random process were true, even the simplest first cell, and all cells thereafter, would not only have needed all the highly complex abilities necessary to live, function, find nourishment, and process that to sustain its own life, but also to reproduce, remember their good qualities, and pass them on with improvements. The offspring would have needed all the above complex abilities, plus the power to survive mutations in order to change into all the other kinds of plants and animals including, eventually, monkeys or their ancestors, ascending into humankind. This first cell or organism would have needed to come into existence with all these highly complex abilities with no intelligent help through a totally unguided happening.

As Sir Fred Hoyle and his associate, Chandra Wickramasinghe, who were Cambridge University mathematicians, cosmologists, and astronomers, have stated, "*The likelihood of the formation of life from inanimate matter is one to a number with 40,000 noughts after it. . . . It is big enough to bury Darwin and the whole theory of evolution.*" [1]

Dr. Stephen C. Meyer, an advocate of intelligent design and co-founder of Discovery Institute's Center for Science and Culture, stated in *Signature in the Cell*, "*The* simplest *extant* (still surviving) *cell, Mycoplasma genitalium – a tiny bacterium that inhabits the human urinary tract – requires 'only' 482 proteins to perform its necessary functions and 562,000 bases of DNA (just under 1,200 base pairs per gene)* ... .

Based upon minimal-complexity experiments, some scientists speculate (but have not demonstrated) that a simple one-cell organism might have been able to survive with as few as 250 to 400 genes." [2] (emphasis added).

Please Note: Using the proper chemical elements and the greatest scientific intelligence and sophisticated equipment available today, our scientists cannot assemble even *one* live, functioning molecular machine for a cell using the known chemical element components.

Knowing these facts, can anyone seriously believe that anything as complex as a whole cell could be assembled into living existence without super-intelligent physical work?

Evolution literally does not do work. By definition it has no intelligence and, therefore, is incapable of performing the enormous amount of intelligent, careful, and precise physical work that is essential to build even the simplest cells, let alone the assembling of these cells into living entities.

Our objective is to provide enough solid evidence for scientific determination that the theory of evolution is now a *falsified theory* regarding the origin and cause of life; it should be disqualified and moved to the history department.

In science, a hypothesis or theory is supposed to be testable for falsifiability as one of its qualifications. The concept that God's work cannot be tested for falsifiability is, itself, false.

Of course, God's role as creator, sustainer, maintainer, repairer, and provider of the breath-of-life to our otherwise inanimate atoms, is falsifiable. We just have to show how all of the intelligent physical work of finding the right numbers of the right atoms in soil, air, and water, then sorting, selecting, counting, grasping, placing, and fastening them in their precise place in each cell of our food, and then redoing all these brilliant steps at lightning speed to build all of our various personal cell parts and cells, with all the decisions and choices this requires, can be performed with no intelligence or guidancc whatsoever.

Considering that with all the intelligence mankind has accumulated, and all the sophisticated equipment we have devised, and that this is totally insufficient to construct even one molecular machine for even one cell, it is plainly obvious that greater intelligence than ours is essential for creating life.

Until it can be shown that this enormous amount of super-intelligent physical work and decision-making can be performed without any intelligence or guidance, Darwinism, Neo-Darwinism, and macroevolution should be set aside as unworkable, falsified, and obsolete theories.

Evolution is not even reasonable, i.e. it does not stand up to the tests of reason.

Charles Darwin himself stated: "*If it could be demonstrated that any complex organ existed which could not possibly have been formed by numerous, successive slight modifications, my theory would absolutely break down."* [3]

Here is the demonstration: *There are many super-intelligent decisions, choices, and physical works essential for the construction of the complex cell machinery and organs of every living creature; evolution simply cannot perform these intelligent works because, by definition, it has no intelligence with which to do the work.*

These super-intelligent works require super-intelligent vision to find, sort, and select the right numbers of the right types of atoms from the wrong ones to make our food and personal cells; to count, grasp, place, and fasten these correct numbers of the correct atoms into their precise positions to create each cell; to install the necessary breath-of-life into the inanimate atoms in these cells to make them live and function; to assemble the right numbers of the right atoms into coded programs necessary to create the DNA and RNA, etc. and make each cell specifically functional; and to do all of this in a sequential fashion at lightning speed.

In recent scientific developments, scientists Jean-Pierre Sauvage, Sir J. Fraser Stoddard, Bernard L Feringa, and the James Tour Group, have shown how great the difficulty is in building and operating even the simplest of molecular machines....totally simplistic compared to any one of the many molecular machines built for us to operate within our cells.

Darwin gave us several other reasons that would show how his theory of evolution could *"absolutely break down."* He knew these potential reasons even before there was a small fraction of today's knowledge of the complexity of life. For example, regarding the complexity of the eye, Darwin states: *"To suppose that the eye with all its inimitable contrivances for adjusting the focus to different distances, for admitting different amounts of light, and for the correction of spherical and chromatic aberration, could have been formed by natural selection, seems, I freely confess, absurd in the highest degree."* [4]

He continues with his theory, however, supposing that from the first, most primitive vision through heredity with modification, the eye could be initiated and improved to be its current status as an "organ of perfection and complication." Yet, *since evolution and natural selection have no intelligence or mechanism to build even one eye cell, or any other cell from atoms in the first place, the theory is, in reality, "absolutely broken down," falsified, and obsolete.*

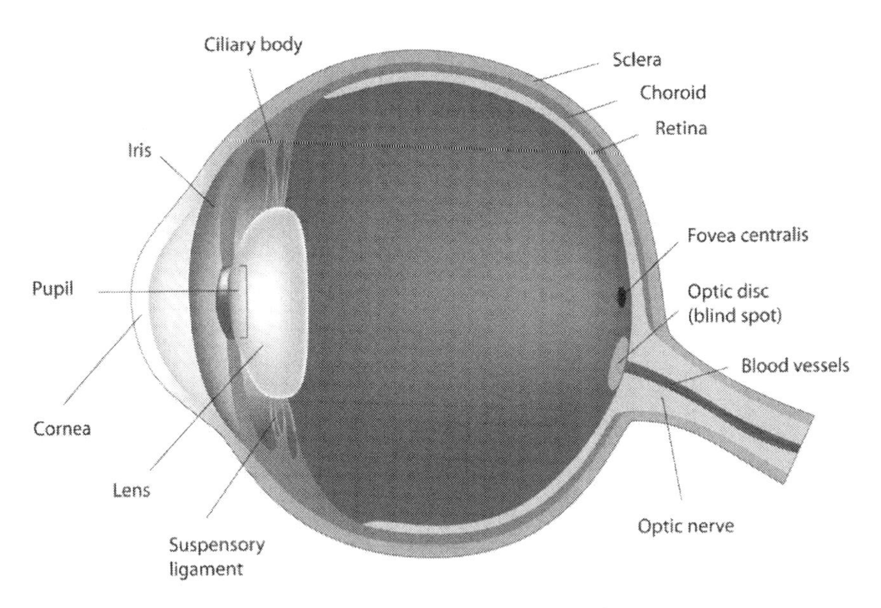

Alila Medical Media/Shutterstock.com

Without today's technology, Darwin could not have fully known exactly how phenomenally complex eyesight is, nor did he understand the super-intelligently brilliant atomic works necessary for the construction, sustenance, maintenance, and repair of every cell involved in our eyesight.

In *Origins,* Darwin says of eyesight: *"[M]ay we not believe that a living optical instrument might thus be formed as superior to one of glass, as the Works of the Creator are to those of man."* [5] (emphasis added).

Obviously, Darwin had great respect for the Creator and His superiority above man's abilities.

Regarding peacock's feathers, Darwin states, "*The sight of a feather in a peacock's tail, whenever I gaze at it, makes me feel sick!"* [6] It seems that the beauty of the design and construction of even one feather in a peacock's tail overwhelmed him.

rickyd/shutterstock.com

If we consider the enormous, brilliant, caring work that goes into designing, constructing, sustaining, and maintaining a peacock's tail-feather, using atoms from the "dust," it *should* be a bit over-whelming to each of us.

Our sincere belief is that this type of example of beauty in creatures and plants must be primarily for God's own enjoyment and ours.

Darwin could not know the details of the enormous physical work involved in building living entities because the technology of the biological construction of cells using atoms was not available to him at that time.

Remember, atoms have no internal means to move themselves into a precise position in a cell, therefore, an external force (with super-intelligent capabilities) is required.

The current rulings against teaching even the possibility of super-intelligent involvement in life, does not change the reality that super-intelligence, vision, dexterity, precision, care, and speed are essential for the life of each living entity.

Ignoring the facts does not change the facts. It is anti-science. We are more than just foolish to continue teaching the false information of Darwinism, Neo-Darwinism, and macroevolution. We cripple our students, future scientists, and leaders by teaching them false information.

It is important to remember that even with all of our accumulated intelligence, specially developed equipment, and chemicals, we and our sciences have not been able to come even a small percentage of the way towards producing even one significant living part of any organism without God's help. Repeating the example, *when using the right chemical elements and our intelligently developed science and equipment, we still cannot build even one live, functioning, molecular machine for a cell.*

Is it not ironic to consider the huge amount of time, intelligence, effort, science, and money going into an endeavour to provide proof that no intelligence is needed to create a living cell?

What this does prove is the exact opposite. It points to proof that super-intelligence IS required to produce living cells and to assemble them into living, functioning plants and creatures.

As scientists have learned more and more about the mind-boggling complexity of even the simplest cells, many are distancing themselves, both formally and informally, from belief in the theory of evolution.

One brave group of scientists have signed a document titled "A Scientific Dissent from Darwinism".[7] There was no compulsion or duress placed upon the scientists who signed this document. The term "brave" is used because many scientists, teachers, and public education institutes (perhaps the majority) are forced or intimidated into teaching only "macroevolution" or "Neo-Darwinism" as the cause of life, by *fear* of termination, penalty, lawsuit, or other unacceptable tactic.

One movie made documenting this disgrace is titled, "*Expelled, No Intelligence Allowed*". [8]

There are examples of other instances including this recent one: A college in Texas was recently intimidated into dropping a non-credit night-school course on the controversy between Evolution and Intelligent Design. Ironically, the names of the intimidating groups portrayed freedom and free thinking in science with input from other science and liberty organizations. [9]

How farcical is that? Their actions in this case went in the absolute opposite direction to the message in their names, i.e. anti-freedom, anti-free thinking, and anti-science, as science is supposed to go where the evidence leads.

Another "liberty" organization is sometimes just as misleading and take their victims, including public education institutions, to court to frustrate the *liberty* to teach in classrooms that there might be some intelligence in the way living entities are designed.

Why are they allowed to get away with such *anti-liberty* actions?

It appears that these organizations (and a few others) are using the 'liberty' of 'free thought' and broad 'freedoms' allowed by constitutional law, to destroy the 'liberty' of 'free thought' and certain 'freedoms' in our educational system.

How anti-American and anti-freedom is that? It is even anti-science!

How have we, in our democratic societies, allowed ourselves to be dictated into this abominable position in our education systems? It goes so much against the beliefs and rights of the majority.

Democracy is gone if a minority is allowed to rule unchecked, especially with our youth.

Good science is supposed to be encouraged to go where the evidence leads. Any actions preventing that are anti-science.

Let's Clarify the Problem as Much as We Can

As Richard Lewontin, an atheistic scientist, stated: *"... for we cannot allow a Divine Foot in the door".* [10] (This seems to be the unwritten motto of the remaining evolutionists).

Well, Richard, you would not be here if you had not been built, sustained, maintained, and cared for by your Divine Creator.

If it can be shown that living entities require intelligent physical work for their design, construction, and life, then Darwinism, Neo-

Darwinism, and macroevolution (which claim to use no intelligence) are falsified as possible causes of life.

isphotography.com/realityrandd.com

That's right, Alex. "Evolution" (Darwinism/macroevolution) *cannot build anything* because it does not have the required intelligence and other necessary capabilities.

By definition, evolution has no intelligence, guidance, appendages, or plan involved. So, how could "evolution" manufacture the right numbers of the right-sized building blocks, paint them, letter them, then grasp onto each proper one and precisely place it to build even this relatively simple, but significant, pyramid message?

Making these wooden building blocks did take some intelligent design and intelligent work, but that was absolutely simple in comparison to the super-intelligent design and physical work involved in building and maintaining each one of our trillions of living cells.

Designing the message took considerable research and contemplation before writing it down and then adding the letters to the properly-made blocks and assembling them into that pyramid message. However, this was truly simplistic in comparison to designing, programming, building, and installing the DNA and RNA molecules for any one tiny living cell.

Another important comparison is that the wooden building-blocks have no life, whereas the inanimate, building-block atoms in each cell *need* to have life "breathed" into them in order for each cell to live and function.

Evolution does not have the super-intelligent ability that is essential to provide the breath-of-life into all the inanimate atoms assembled into our cells. Also, it can't deliver the cells to their precise place, and it can't fasten them there.

To build us, *super-intelligent vision, dexterity, intelligence, speed, precision, and care* are needed to find, sort, count, grasp, precisely assemble, and fasten the right numbers of the right specific atoms required to build each one of the various cells of our foods and then our bodies.

Evolution, by definition, *cannot do this*, as it has no brain or intelligence, no super-intelligent vision to see the right atoms and decide where to put them, no appendages to grasp the atoms and place them, and no plan, no speed, and no care to precisely assemble, place, and fasten the right numbers of the right atoms from our foods into precise places in different cells for the various parts of our body. Therefore, the Theory of Evolution as the cause of life is falsified.

These are just a few of the reasons why a super-intelligent force with these super-capabilities, is required to do all of the brilliant, reliable, caring, physical work needed to cause life.

In an attempt to understand where the prompting for the anti-Creator thoughts are coming from, we can look at Chapter 7 for some clues: - some people do not like their concept of God or Creator because they have been hurt or offended by someone pretending to be a Godly person; -some have been subtly deceived by the other super-intelligent force who destroys through temptations; -some mistakenly think that accepting God will take away their freedom. (Actually, the opposite is true); -some look at God as an intruder into their life (not realizing that He is their best friend as no one works harder for them without even receiving a "Thank-you").

Two major points we must remember are:

1. We cannot blame God for what people do wrong; and
2. No one works harder or cares more for each one of us than our Creator God.

He builds into us the freedom to make our own choices, good or bad.

For good reasons, our four nations highlighted is this primer, were founded on Godly principles. The majority of the citizens in these countries continue to believe in Him, according to census after census. Some of the reasons they believe in God include their appreciation for the following:

- for the food that is made for all of us (whether we thank Him or not);
- for healing of their wounds;
- for answered prayers;
- for His guidance and wisdom;
- for the family and friends that God has made for them;

- for the beauty in the flowers, trees, pets, birds, and other creatures made for our enjoyment; and so on.

It seems that the majority of citizens do not have enough faith to believe like atheists who think that all this work is done for them by some random, unintelligent, accidental, magical, or unguided process.

Here is another message with youth involved; a song named "Goodbye Evolution," sung by "Karen and the Kids":

"Satan uses Darwin's theory to make folks forget God,
But he can't fool all the people all the time.
God still faithfully makes our food from dust and rain,
And He makes our cells from our food; it's all Divine.
Evolution is the worst hoax of all time!
Oh, everything's made of atoms, and atoms don't have legs.
Mr. Darwin did not know this, but atoms don't have legs.
So, someone has to place each atom to build each living thing.
Atoms cannot jump to their right place; someone creates live things.
It really is so simple; the scientists could agree,
That everything's made of atoms, and atoms cannot see.
So, someone has to place each atom to build each living thing.
Atoms cannot jump to their right place; someone creates live things.
It has to be someone real smart 'cause I'm no piece of junk.
My trillions of cells each play their part; my builder is no punk.
Yes, someone has to place each atom to build each living thing.
Atoms cannot jump to their right place; someone creates live things.
So, goodbye evolution; to most you make no sense;
Except to population who haven't seen love immense.

But no one ever loves them more nor works so hard for them,
Than God their great Creator; all life depends on Him!
All life depends on Him! On Him!
GOODBYE EVOLUTION !!" [11]
(available on iTunes)

Where Do Our Building-block Atoms Come From?

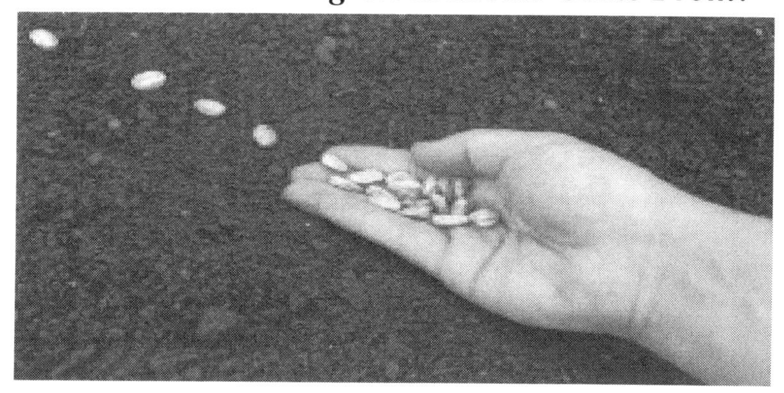

Azure/Shutterstock.com

All the building-block atoms required to build you and me are right here in the "dust" (soil), rain, air, seeds, and sunshine. The garden or field is like a shopping mall where the right numbers of the right building materials (the right elemental atoms) have to be found amongst all the unneeded ones, then carried to the building site for assembly into our fruit and veggies.

All else that is needed is the brilliant super-intelligent force that can find all the right numbers of the right atoms and assemble them exactly in accordance with each seed's requirements, into our various foods.

In the simplistic diagram below, the colored dots represent various elemental atoms in the soil of a garden. They are next to an enlarged image of a section of a carrot seed planted in the soil.

You have to use your imagination here because it would be dark inside the soil and dark inside the carrot seed. The seed may have been dormant in a package for three years before being planted. If we assume the gardener has put some water here because she wants the carrot seed to start "growing", how do these atoms now start being moved to their needed position? They need to be assembled in rapid sequence, to build the cells for the first little root hairs, precisely and properly attached to the seed in order to begin the carrot building process. If you were the builder, where would you decide to start building the root hair and which atoms would you choose, (and which would you decide to leave behind), and in what sequence?

This example is just to give a little more graphic detail of the super-intelligent work necessary to start the construction of some of our food.

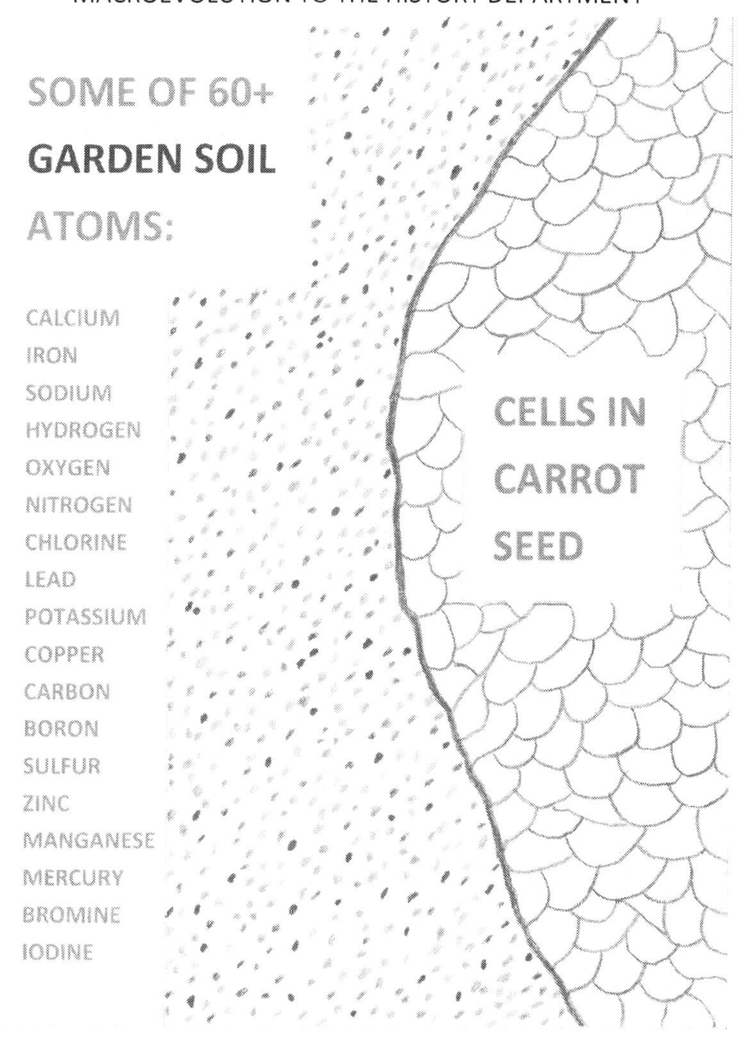

Reality R&D

So, here is the portrayal of atoms in the soil around a carrot seed. Can you imagine the sequential physical work-steps it will take to build the carrot? Only the right numbers of the right atoms will be selected and precisely placed to build even the first tiny root hairs. The wrong atoms will not be chosen or precisely placed. Do you see the need for super-intelligence to make the plan, decisions, choices, and accurate placement of atoms for this construction project?

Our body is normally constructed of about 59 or 60 types of elemental atoms. Besides the elements shown in the diagram, we need food containing Phosphorous, Magnesium, Fluorine, Silicon, Rubidium, Strontium, Aluminium, Cadmium, Cerium, Barium, Tin, Selenium, Nickel, Chromium, Arsenic, Caesium, Molybdenum, Germanium, Cobalt, Antimony, Silver, Niobium, Zirconium, Lanthanum, Tellurium, Gallium, Vitrium, Bismuth, Thallium, Indium, Gold, Scandium, Tantalum, Vanadium, Thorium, Uranium, Samarium, Tungsden, Berylium, and Radium. Oxygen, Carbon, Hydrogen, Nitrogen, Calcium, Phosphorous, Potassium, Sulfur, Sodium, Chlorine, and Magnesium are the primary components by volume in our bodies. The rest are in just trace proportions but still important to our health.

That process of finding and assembling the right numbers of the right atoms from our eaten foods has to be repeated to build the great variety of new cells as needed for *our* body and life.

When you understand even just a little about engineering, construction, and manufacturing, as well as maintenance, repair, and replacement of damaged parts, you know the importance of assembling the right materials accurately in accordance with good design plans. Precisely assembling, placing, fastening, and programming the functional codes for all of our highly complex cells is absolutely a matter of life and death.

It is enormous, brilliant, reliable, caring work performed at lightning speed for every one of us. This includes atheists, evolutionists, and others who do not yet acknowledge this super-intelligent force that works so hard and cares so much for each one of us, every second of every day.

This two-step process of making us from the "dust" is essential for our construction, sustenance, growth, maintenance, and repair. All of this amazing work cannot be performed in an unguided, random manner or left to chance.

Think of just one of our marvelously designed and built organs, like our heart, for example. We know that if we live for 80 years, it will have to be well enough designed, constructed and maintained to do the following: it has to 'beat' or 'pump our blood' about 60 times per minute x 60 minutes/hour = 3.600 times per hour x 24 hours/day = 86,400 times per day x 365 days/year = 31,536,000 times per year x 80 years = 2,522,880,000 heart-beats or pumps in our first 80 years.

How brilliantly made and carefully maintained is that piece of human machinery, especially if we do our part to help ourselves? And how about all our other organs that work so well for so long. Can we give a little "Thank you" to God for His care for us?

An enormous amount of brilliant work is provided to us every day. Why would it be so reliable and constant if the motivation to provide it were not immense, loving care for us? For what other reason could it be?

Since all of this necessary work and care and much more are provided to us free of charge for our whole lifetime, it is easily understandable why we can all rightfully say, "In God We Can Trust".

No wonder many of the founders and leaders of our four nations recognized, acknowledged and consulted Him and have made Him an official part of our governance. And why would we keep it virtually illegal to teach our students anything about Him?

We have summarized a number of the key factors in the "*...absolute breakdown of the theory*" (of evolution), to use Mr. Darwin's own words.

The following seven basic principles of life provide some of the reasons why a phenomenally intelligent, hard-working, immensely reliable, trustworthy, and caring super-intelligent force is required to design, construct, sustain, maintain, and repair all living entities, especially us human beings.

Seven Basic Principles of Life

Principle #*1*

Virtually all matter, living or not living, is constructed or built of atoms that do not have legs, brains, fins, or muscles.

Since atoms have no internal means to move themselves into any precise position in a cell, a super-intelligent and capable external force is required to find, sort, select, count, grasp, and precisely place all of the right numbers of the right atoms necessary to build each cell of our food (fruit, vegetables, etc.); then, to perform a similar process with the right numbers of the right atoms from our digesting food and place them precisely to build each new complex cell for us; and then deliver each one of these cells to its precise position in our body, fasten it there, and hook it up properly to our blood vessels and nerve networks.

Every step of this brilliant work must be performed with super-intelligent planning, care, dexterity, precision, and speed. In the English language, this phenomenally intelligent, reliable, trustworthy, and caring super-intelligent force is called "God".

(As Evolution, by definition, has no intelligence to work with, it is falsified as the cause of life).

Principle #*2*

The super-intelligent physical work required to build living cells, goes far beyond the parameters of possibility for the unguided process theorized as Darwinism, Neo-Darwinism, or macro-evolution.

For example, over 4900 quadrillion right atoms per second must be found and sorted from our digesting food, then selected, precisely assembled into new red blood cells, and delivered to each average adult's blood stream just to supply his or her replacement red blood cells; that is for every adult every second of every day.[12]

Of course, in the same second, a greater number of atoms (to make the leaves, roots, skin, etc. for our fruit and veggies) have to be found, sorted, and selected from the soil, water, and air of fields, gardens and orchards to be assembled into more food from which to take the right numbers of the right atoms to make each future second's red blood cells for each of us.

This is in addition to the phenomenal, reliable, caring physical work of constantly maintaining, repairing, and/or replacing the other approximately 80 trillion cells in each human adult body.

Evolution, by definition, has no intelligence to use, therefore, *it is incapable of doing any of the super-intelligent physical work necessary to build each cell.* (Thus, Evolution is falsified as the cause of life).

Principle #*3*

"Dead dogs don't bark," which is to say that although all the right atoms, molecules, and cells are precisely built and placed into their correct position for its eyes, ears, teeth, brain, legs, heart, lungs, paws, liver, kidneys, stomach, fir, claws, nose, and so on, *without the super-intelligent breath-of-life, those atoms, molecules and cells are_not going to move one millimeter.* This God-given breath-of-life is crucial to every living entity. When it is removed, the entity's life ends. (Evolution cannot provide this necessary attribute, therefore its theory as the cause of life, is falsified).

Principle #*4*

Evolution is not a force. By definition, it has absolutely no intelligence, guidance, appendage, or plan to work with. *However, all of these attributes plus super-intelligent vision, dexterity, precision, and speed are essential to construct, grow, maintain, repair, and care for each living entity.* (As Evolution is not capable of performing these tasks, it cannot be the cause of life, and is therefore falsified as the cause of life).

Principle #*5*

Mr. Darwin said his theory of evolution could "absolutely break down" if the fossil recorders did not find examples of one kind of creature changing into another kind of creature. This would indicate that creatures do not evolve from one kind to another kind. He knew there would need to be huge numbers of transition fossils. In his book, *On the Origin of Species,* he states, *"But just in proportion as this process of extermination has acted on an enormous scale, so must the number of intermediate varieties, which have formerly existed, be truly enormous. Why then is not every geological formation and every stratum full of such intermediate links? Geology assuredly does not reveal any such finely-graduated organic chain; and this, perhaps is the most obvious and serious objection which can be urged against the theory (of evolution)."* [13]

Out of the millions of fossils found, there should be hundreds of thousands of transitions, but they do not exist. For example, there are no complete fossils of a man with a monkey's tail. Men and monkeys still exist, yet there is not even one complete transitional creature in existence or in the fossil record.

What we have found are minor changes within species such as different colors and sizes of humans, horses, cats, dogs, fish, birds, bugs, etc.

The occasional random mutations occurring between generations, that were at one time thought to be the cause of improvements in a species, actually involve a loss of DNA information. A more accurate word for these mutational changes is "devolution" or "degeneration," *never a change of one kind of entity to another kind.*

There have been many attempts to produce images of transitions from bone fragments and there are fossils that have some characteristics similar to other kinds of creatures. However, the understanding of the enormous amount of intelligent work involved in designing and building living entities with atoms, points far more logically to a common designer and builder rather than a common ancestor.

As stated by Dr. Gary Parker, a paleontologist, biologist, and former evolutionist, *"Fossils are a great embarrassment to Evolutionary theory and strong support for the concept of Creation."* [14] (The lack of transitional specimens is a reassurance that Evolution is not the cause of life).

Principle #6

All living entities are built of cells containing DNA. DNA is like a computer software program which has very sophisticated, highly complex coding to assist in the multiple and varied functions which cells have to perform in order to help keep a living entity alive.

Complex, intelligent, functional codes, like those in DNA, cannot be programmed without an intelligent programmer. (Therefore, Evolution is falsified as the cause of life).

Principle #7

There are many factors that must be intelligently tuned, highly regulated, and crucially consistent in order for our beautiful planet to function, and for living entities to exist. Random, uncontrolled, inconsistent conditions would quickly lead to extinction of creatures.

Here are just a few conditions under which life would not exist:

1. If temperatures became too hot or too cold;
2. If there was insufficient water available;
3. If there was insufficient food;
4. If there was no one to build cells from atoms;
5. If there was no one to breathe life into these cells;
6. If there was no controlled sunlight;
7. If there was no controlled atmosphere;
8. If there was no controlled gravity;
9. If there was no controlled electricity;
10. If there was no super-intelligent force to keep all of these necessary factors (and more) in balance, life would not exist.

Fortunately for the creatures on our planet, we have a super-intelligent controller who reliably and consistently keeps everything necessary for life in balance. (As Evolution, by definition, has no intelligence to work with, it is falsified as the cause of life).

Here is a summary list of e*ssential, intelligent, physical works* required to build and maintain each living entity that *evolution cannot perform, as, by definition, it has no intelligence:*

1. Evolution cannot *see and think*;
2. It cannot *find* the necessary 'building-block' atoms;
3. It cannot *sort* the right atoms from the wrong ones for building each cell-part;
4. It cannot *count* the right numbers of each type of atom for building each cell-part;
5. It cannot *grasp and move* the right atoms for building each cell-part.
6. It cannot *precisely place* each atom to build each cell-part;

7. It cannot *fasten* each atom in its correct place in each cell-part;
8. It cannot *program* the RNA and DNA molecules as required for each cell;
9. It cannot *breathe life into* the inanimate atoms in each cell;
10. It cannot *re-sort* the atoms in eaten food to build our human cells;
11. It cannot *work quickly,* or at all;
12. It cannot *build* a living entity;
13. It cannot *sustain* the life of a living entity;
14. It cannot *maintain* any entity;
15. It cannot *repair* or heal any entity;
16. It cannot *build communication systems* within living entities;
17. It cannot *fine-tune our living conditions* on our planet;
18. It cannot *build blood-vessel networks* for creatures.

These are only a few of the super-intelligent works necessary for life on our planet, that evolution cannot perform. Therefore, Evolution is falsified as the cause of life.

At the time of the founding of our nations, not all of this science was known or understood in detail by our founding fathers.

However, the need for super-intelligent external help with: growing crops of grains, vegetables, and fruit for food; the producing of cattle for milk and meat; fish and fowl to eat; etc. was all appreciated and understood to be super-intelligently, reliable, and faithfully created for the benefit of mankind.

They called this super-intelligent external provider, "God", just as the majority of citizens in our four nations do today.

It is time to teach the truth of *essential super-intelligence, physical works, and awesome care* for the design, construction, sustenance, maintenance, and repair of all living entities, especially us.

What could be more obvious than the need to provide *accurate* information for our students of all ages and stages of their education?

How can we continue to teach our future problem-solvers false information, like Evolution as the cause of life, then expect them to come up with the best solutions?

If we teach them that the essentials of life just *randomly or magically* happen, and then expect them to develop the best food plans, medicines, and so on, how bright is that on our part?

For some people, evolution may have seemed like a desirable explanation as the cause for and development of living entities. However, in light of recent discoveries in science, *essential, intelligent, physical work* for the design, construction, maintenance, and repair of highly complex living cells and entities should become, by far, *the best explanation of the cause of life.*

What we can thank evolutionists for is their thorough work in attempting to prove through all possible means that there is no need for a Creator. It seems that they have left no stone unturned in endeavoring to show another possible way that living entities could have been designed, built, sustained, maintained, and repaired without any intelligent help. *As they have performed this exhaustive search for an alternative to our Creator, and have failed, it is, therefore, with greater ease that we can lay this Theory of Evolution to rest as an unworkable theory for explaining the cause and complexities of Life.*

There are no more arrows in their quiver.

God's Style of Evolution: – evolving a crawling worm into a beautiful flying butterfly in a matter of days, not millions of years.

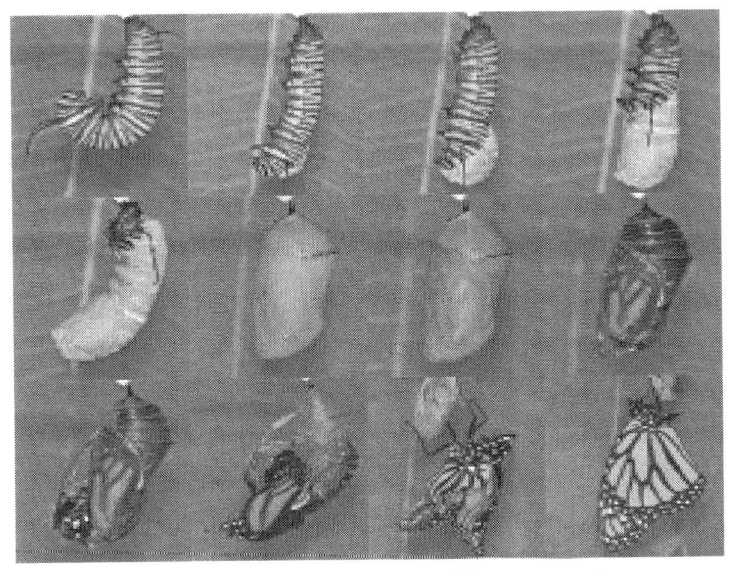

Stephen Russell Smith Photos/Shutterstock.com

Does *anyone* seriously believe that there is no intelligent work involved in this magnificent and speedy demolition and reconstruction work?

It is easy to understand why there needs to be an enormous amount of super-intelligent, dexterous, skillful, careful, precise, essential, physical work involved that evolution, by definition, cannot perform.

Monarch Butterfly migration is another awesome phenomenon requiring super-intelligent help.

The Scientific Method for Verification of Factualness

"To be considered a scientific 'fact'," says Dr. Kenneth L. Currie of the Department of Geological Survey of Canada, *"a concept must pass two tests. First, it must be repeatable; that is, the same concatenation of circumstances always produces the same result. Second, it must be impersonal; that is, the same set of circumstances yields the same observation to different observers, or to a suitably adjusted machine."* [15]

Similarly, as stated by Gary Kemper, Hallie Kemper, and Casey Luskin, *"The scientific method* (for verification of a scientific fact) *is usually described as a four-step process consisting of: – 1. Observation; – 2. Hypothesis; – 3. Experimentation; – 4. Conclusion."* [16]

Using the Scientific Method, let's start verifying some examples of God's phenomenal work for us through this potentially new branch of Godly life-science we are calling "Atomic Biology".

VERIFICATION #1: The super-intelligent force known as our Creator God, constructs our vegetables and other food plants.

Step 1. Observation

To observe the construction of plants that we can use for our own nutrition, let's choose a carrot seed, a cucumber seed, a strawberry seed, and a wheat seed. Let's put them into an open-top glass box, like an aquarium, after we fill it three-quarters full of garden soil. Let's put the seeds right against the front glass about one inch from the top of the soil, so we can observe their growth.

The various kinds of seeds that God creates contain the brilliant, basic design plans for their kind of living entity.

Each and every time we observe the "growth" of any entity, we are witnessing "construction" in progress. In the case of a building, all the materials plus builders' work and skills are visible. In the case of living entities for our food, the building materials are the right numbers of the right atoms from the soil, water, and air; although, they and the skills to handle them are too minute to be seen (at this time), the results of the construction work are visible.

Let's add a little water to the soil and watch what happens daily over the next few weeks. Use of a good magnifying glass will make it more interesting. Anyone can do this.

If we want to get sophisticated, we could use a magnifying video camera that takes a picture about every fifteen minutes to record the construction of each type of food plant.

We almost don't have to go through this process because we know that the Builder is so reliable and has been throughout the history of humankind. We can trust Him and rely on Him to construct these food items for us, just as our ancestors have for millennia.

We will observe the gradual assembly of atoms into tiny root hairs being built onto each seed using atoms from the soil, water, air, and some from the seed. Although we are not able to actually see the individual atoms being sorted, selected, grasped, and precisely placed in each cell, we know this has to be happening by seeing the *results* of this intelligent, physical work, – construction progress, and the finished products, complete with the vitamins, minerals and other nutrients built in for us.

The majority of the atoms in our foods and bodies include carbon, hydrogen, nitrogen, oxygen, sodium, calcium, iron, potassium, phosphorous, and smaller quantities of a large variety of other elemental atoms.

In a few days, we observe the construction of the bodies or stems of each plant. The right numbers of the right building-block atoms are being found at the right time in the soil and water, then sorted, selected, grasped, conducted in through the roots, and precisely assembled into the right type of cells needed for the next steps in the plant's progressive, critical-path construction.

Various colors are added to the parts of the plants as the builder desires, and always some green for the chlorophyll that helps to gather energy and other aid from the sunlight, or a similar artificial light.

We can observe each food plant being reliably built for us according to its seed type and design, right up to maturity, complete with more seeds. Carrots are usually harvested before their seeds are manufactured, but some are allowed to go on to the seed production stage.

A single wheat stock can be built with 20, 40, or more seeds that are normally made into cereal or flour for our bread or other baked goods.

We can repeat endlessly the planting of seeds and the growing of food plants, as others have for millennia, with the same life-sustaining nutrients built-in by the Builder.

Step *2*. Hypothesis

We know that *atoms have no internal means or intelligence to move themselves precisely* into the correct place in each cell, therefore *they require a super-intelligent external force* to grasp the right numbers of the right atoms, move them, assemble them quickly, and fasten them precisely and reliably according to an extremely complex and sophisticated plan.

Remembering that evolution has no intelligence with which to operate, and scientists have shown that with all our human intelligence and sophisticated equipment, we cannot come anywhere close to constructing even one small, living, functional molecular machine for a cell, we have thus shown that a *greater intelligence than we have is required to design and build living entities.*

Because of the super-intelligence required by this external force, the best explanation is to credit our Creator, the God of our nations, for this phenomenal, life-giving, physical work.

Step *3*. Experimentation

We can experiment over and over again by planting a huge variety of seeds, from tiny mustard seeds to larger watermelon seeds, to grow our foods.

The planting and growing of seeds can be repeated endlessly as has been performed for millennia with the same reliable, life-sustaining nourishment included by the Builder.

We do not need to do this in a laboratory as we can see this amazing construction work going on in every garden, field, orchard, forest, lake-bottom, or sea-bed.

One advantage of planting seeds in a glass case is that we can observe the construction of each type of plant, starting with its first tiny roots.

Imagine all the other work needed to build enough food for all seven billion people on Earth every day. God does build enough for everyone, but we don't share it with others as well as we could and should.

In the early 1990's, when government debts were soaring up, I was working on another project to find ways of reducing government expense. One of my relatives in South Africa sent me some special information that they gave to their poor and needy people instead of money. It was one double-sided page of instructions on how to grow enough food for one person, on a piece of ground the size of a door (3 feet by 6 feet)! Diagrams showed how to plant various seeds, harvest the vegetables and fruit when ready, return leaves and peelings to the ground along with more seeds, and keep rotating the process. This works most easily in climates with a year-round growing season, but with suitable green houses, growing time can be extended anywhere.

If you fly over land for five minutes from any of our airports, you can generally see huge acreages of vacant land with no crops growing. Therefore, food production could be increased immensely.

Building veggies and fruit using the atoms from the dust, air, and water is relatively simple in comparison to building one of us, but some parts of the work are similar.

Remember that just for our red blood cell replacement, *quadrillions* of the right atoms have to be found and precisely placed *every second of every day* for each one of us.

As with plants, we started out as seeds, but in our mother's womb. The two combined seeds began to grow and be built into us by the precise placement of all the right numbers of the right atoms, which had to be found in the food our mother had eaten each day. And of course, these atoms came from the soil, air, and water that were the source of the atoms as building-blocks for her food.

Experiments that involve planting various kinds of seeds and observing them being built precisely and carefully using the right atoms from the soil, air, and water, into each respective kind of plants, <u>are repeatable and impersonal</u> – the two requirements for verifying factualness.

Experiments involving the building of various creatures from their seeds, using the right atoms from their foods, are more difficult to observe, but with the help of ultra-sound and other observation equipment, we know what is happening. These experiments are also repeatable and impersonal.

Step *4*. Conclusion

The best explanation for the origin and development of living cells and entities is that super-intelligent, physical work is essential to design, construct, sustain, and maintain each one.

VERIFICATION #2: No construction of food items is possible without the super-intelligent force known as Creator God and the seeds He has designed and built.

Step 1. Observation

If we want to try to observe the construction of some food-plants *without God's help*, we cannot use seeds because we know that they require His super-intelligent design and dexterous, precise, construction work to exist. These are capabilities that Darwinism, Neo-Darwinism, and macroevolution, by definition, *do not possess*. In fact, even with the best scientific intelligence that humanity can provide, we have been shown that not even one tiny living functional molecular machine for a cell can be built without God's help, let alone an entire seed.

Let's use the same glass container and soil from Verification #1. Just observing the opposite side of the container of soil will make it easy to duplicate the conditions and available materials with the exception of the God-built seeds.

We have to take into account that there are millions of microbes (microscopic living entities) in every cubic foot of garden soil; therefore, birth, growth, and life are created and cycling constantly. There may even be some seeds for some kinds of plant life in the soil that God will begin building into their kind of entity when we add water. However, rather than sterilizing the soil or just using the necessary pure elements in sufficient quantities and being criticized for rigging the experiments, we are using common garden soil.

We do this experiment, and the previous one, in the same glass case simultaneously so that the soil, weather conditions, and air supply are identical.

As expected, where the God-built seeds were planted and watered, we observed the exact kinds of food-plants being built as the seeds that were planted.

Where there were no food seeds planted, even though the same soil ingredients and water for building the same kinds of food plants existed, no food plants grew.

This observation can be witnessed in gardens, fields, and orchards around the world with constant repetition and reliability for all races of people. Food plants only grow where their seeds are planted, formally or informally.

Step *2*. Hypothesis

Regardless of how many times this experiment is repeated and by whom, no food-plants will ever grow unless God-built seeds are planted, formally or informally. Seeds may be *altered* through tampering by scientists, but not *created* by them.

Step *3*. Experimentation

Even using the best, clean garden soil, the best fertilizers, and the best water and air available, no matter how carefully the fertilizing and watering is done, no food plants are built without the planting of their seeds.

Step *4*. Conclusion

We believe it is safe and logical to conclude that regardless of how many times this experiment is repeated and by whomsoever, no food-plants will ever 'grow' unless God-built seeds are planted. This satisfies the tests of repeatability and impersonal results, i.e. the same result over and over for any observer. It also passes the test of reason.

Our accumulated research regarding the phenomenal amount of super-intelligent physical work that must be performed to build, sustain, and maintain any food plant, and any one of us, is one of the most compelling rationales that eliminates the possibility of an unguided or unintelligent process as the designer, builder, sustainer, and maintainer of living entities.

Until proven otherwise, the best explanation for the origin and cause of life in all living entities appears to be the enormous amount of super-intelligent, physical work required for the design, construction, and ongoing sustenance and maintenance of each cell in each living entity. As it is essential that all of this work be performed with super-intelligence, super-dexterity, precision, care, reliability, and speed, it follows that a super-intelligent force with all these capabilities, must be performing all of this work.

Quotable Quotes from Significant Scientists:

1. *"Is it really credible that random processes could have constructed a reality, the smallest element of which – a functional protein or gene – is complex beyond...anything produced by the intelligence of man?"* [17]

Michael Denton, biological research scientist and M.D.

2. "*At that moment, when the RNA/DNA system became understood, the debate between Evolutionists and Creationists should have come to a screeching halt.*" [18]

I. L. Cohen, archaeologist and mathematician.

3. "*The likelihood of the formation of life from inanimate matter* (without super-intelligent help) *is one in a number with 40,000 noughts* (zeroes) *after it...It is big enough to bury Darwin and the whole theory of evolution.*" [19]

Sir Fred Hoyle and Chandra Wickramisinghe, astronomers, cosmologists, and mathematicians.

4. "*An intelligible communication via radio signal from some distant galaxy would be widely hailed as evidence of an intelligent source. Why then doesn't the message sequence on the DNA molecule also constitute prima facie evidence for an intelligent source?*" [20]

Charles B. Thaxton, Walter L. Bradley, Roger L Olsen, chemists.

5. "*That organic evolution could account for the complex forms of life in the past and the present has long since been abandoned by men who grasp the importance of the DNA genetic code.*" [21]

John Grebe, chemist.

6. "*From the claims made for neo-Darwinism one could easily get the impression that it has made great progress towards explaining Evolution... In fact, quite the reverse is true.*" [22]

Peter T. Saunders, mathematician. and Mae-Wan Ho, geneticist.

7. "*We have to admit that there is nothing in the geological records that runs contrary to the view of the conservative creationists, that God created each species ...*" [23]

Edmund Ambrose, evolutionist, cell biologist.

8. "*There is no recorded experiment in the history of science that contradicts the second law (of thermodynamics) or its corollaries...*". [24]

G. Hatsopoulous and E. Gyftopoulos, physicists.

(The second law of thermodynamics means, in part, that everything left unmaintained by an intelligent force, tends to deteriorate, not improve, with the passage of time).

9 "*Of all the statements that have been made with respect to theories on the origin of life, the statement that the Second Law of Thermodynamics poses no problem for an evolutionary origin of life is the most absurd....*" [25]

Duane Gish, biochemist.

10. "*The Evolutionist thesis has become more stringently unthinkable than ever before...*" [26]

Wolfgang Smith, physicist and mathematician.

11. "*As is now well known, most fossil species appear instantaneously in the fossil record.*" [27]

Tom Kemp, curator of Zoological Collections at Oxford U.

12. "*Evolution requires intermediate forms between species and paleontology does not provide them.*" [28]

David Kitts, paleontologist and evolutionist.

13. "*The preservation of numerous soft-bodied Cambrian animals as well as Precambrian embryos and microorganisms undermines the idea of an extensive period of undetected soft-bodied evolution. In addition, the claim that exclusively soft-bodied ancestors preceded the hard-bodied Cambrian forms remains anatomically implausible.*" [29]

Stephen C. Meyer, science philosopher.

14. "*A growing number of respectable scientists are defecting from the evolutionist camp, moreover, for the most part these "experts" have abandoned Darwinism, not on the basis of religious faith or biblical persuasions, but on strictly scientific grounds....*" [30]

Wolfgang Smith, physicist and mathematician.

15. "*It is the sheer universality of perfection, the fact that everywhere we look, to whatever depth we look, we find an elegance and ingenuity of an absolute transcending quality which so mitigates against the idea of chance.*" [31]

Michael Denton, biological research scientist and M.D.

16. "*The theory of evolution...will be one of the great jokes in the history books of the future. Posterity will marvel that so flimsy and dubious an hypothesis could be accepted with the incredible credulity that it has.*" [32]

Malcolm Muggeridge, philosopher.

17. *"Human DNA is like a computer program but far, far more advanced than any software we've ever created."* [33]

Bill Gates, founder of Microsoft.

18. *"...the amount of information that could be stored in a pinhead's volume of DNA is staggering. ... a pinhead of DNA would have the equivalent information of a pile of CD's 1,000 miles high, or 40 million times as much as a 100-gigabyte hard drive."* [34]

Jonathan Sarfati, physical chemist.

19. *"No undirected physical or chemical process has ever demonstrated the capacity to produce specified information starting from purely physical or chemical precursors. For this reason, chemical evolutionary theories have failed to solve the mystery of the origin of first life—a claim that few mainstream evolutionary theorists now dispute."* [35]

Stephen C. Meyer, science philosopher.

20. *"It is certainly true that a machine carefully made by a craftsman reflects the existence of its creator. It would be foolish to suggest that time and chance could make a computer or a microwave oven, or that the individual parts could form themselves into these complex mechanisms.... Yet life is far, far more complex than any man-made machine."* [36]

Paul S. Taylor, science author and motion picture producer.

21. *"Any suppression which undermines and destroys that very foundation on which scientific methodology and research was erected, evolutionist or otherwise, cannot and must not be allowed to flourish. ...It is a confrontation between scientific objectivity and ingrained prejudice. ...In the final analysis, objective scientific analysis has to prevail – no matter what the final result is – no matter how many time-honored idols have to be discarded in the process....*

    It is not the duty of science to defend the theory of evolution, and stick by it to the bitter end – no matter what illogical and unsupported conclusions it offers... If in the process of impartial scientific logic, they find that creation by outside super-intelligence is the solution to our quandary, then let's cut the umbilical cord that tied us down to Darwin for such a long time. It is choking us and holding us back.

    ...Every single concept advanced by the theory of evolution (and amended thereafter) is imaginary as it is not supported by the scientifically established facts of microbiology, fossils, and mathematical probability concepts. Darwin was wrong. ... The theory of evolution may be the worst mistake made in science."[37]

    I. L. Cohen, archaeologist and mathematician.

We now know that there is an enormous amount of super-intelligent physical work essential for building each living cell and entity – work that evolution, by definition, is incapable of performing as it has no intelligence and does no work.

If the theory of evolution cannot provide an explanation of how it finds the right atoms in the soil, air, and water to construct all the cell-parts required to build any vegetable or fruit; how it sorts the right atoms from the wrong ones; how it counts the right numbers to select; grasps these right numbers of right atoms and places them precisely and fastens them securely and properly to build each cell-part; how it programs the DNA and RNA, how it builds the molecular machines, and other crucial parts in each cell; how it adds the essential breath-of-life to make each cell live and function properly; and how it can precisely handle over 4900 quadrillion right atoms per second in each adult, just to build his or her replacement red blood cells as just one part of our life support; and how this can all be done with no intelligence or guidance, then the theory of evolution should be fully classified as falsified.

If Darwinism/Neo-Darwinism/macroevolution do not work, are incorrect, falsified, and misleading, by what justification should they be taught?

This is a serious question for our scientists, for our educators, for our leaders, for parents, and for students.

We all receive an immense amount of God's physical work and care every second of every day.

It is time to make the change and bring understanding of our Creator, and God of our governments, back to our classrooms.

Chapter 6
Choices and Consequences

In this primer, we are looking at the USA, the UK, Australia, and Canada because it is historically noted that our Creator God provided a significant part of the foundational wisdom involved in the establishment and operation of these four nations. He continues to be a significant part, although it appears that a smaller portion of our students, scientists, citizens, business leaders, and politicians are seeking His advice and guidance for their lives and work today.

As every structure is best served by a good, strong foundation, so it is with nations.

Through the decades, the beneficial, God-based operating philosophy of these four nations has attracted many immigrants from virtually all other countries of the world, which do not have as attractive a governing system.

To what can we attribute the initial wisdom, strength, justice, integrity, stimulating atmosphere, attractiveness, and balanced growth of these four nations? – Certainly, when you look at the declarations, constitutions, and other founding documents and laws of these nations, it is easy to see the Godly principles adopted to guide our governments.

As with buildings, it is necessary that the whole structure is solidly *built, maintained, and repaired promptly when necessary.*

Probably, the worst choice made by governments since 1960 was this:

In the early 1960s a lady by the name of Madalyn Murray (O'Hair), an outspoken atheist with some atheist friends, influenced the US Supreme Court members to choose to declare that prayer and Bible reading be banned in US public schools.

The members must not have realized how drastic the consequences for society would be from that terrible choice.

Similar policies were adopted in the UK, Australia, and Canada.

In the following years through today, there has been an appalling increase in: drug and alcohol abuse, sexually transmitted diseases, and births to unwed mothers (according to stats from the US Department of Health and Human Services); in violent behaviors, and single parent families (according to the US Department of Commerce, Census Bureau); and a major drop in academic achievement (according to the College Entrance Exam Board, New York).

The resulting emotional and physical devastations have led to further personal abuses, deep depressions, and increased suicides. What a terrible choice that was, in the early 1960s. The consequences have been absolutely abominable for our students, citizens, and nations.

What are the causes and how do we reverse this downward trend?

At about the same time as Bible readings and prayer were removed from classrooms, the misleading teaching of Darwinisms and evolution-only as the cause of life, began to be enforced in our secular schools and universities.

This was another terrible choice with widespread harmful consequences as it removed the teaching of understanding the individual care for each student by their Creator.

What a detrimental state for education to have descended into. It is totally anti-science and anti-society.

Fortunately, thinking scientists and educators are now choosing to back away from evolution as the cause of life, as much as they are allowed without losing their jobs.

This is a beginning of restoration and recovery.

Nations today can deteriorate and collapse when they allow destructive choices to grow within them. They are like forms of national cancer and they are not new. We saw examples in the twentieth century and many more in the old testament and other history books.

This is not religion. It is pure common sense supported by history.

Since belief in Evolution as the cause of life, knowing how brilliant and complex life is, requires immense Faith, this makes Evolution as much of a Religion as any other Belief requiring Faith.

If we look at the troubled areas of our world today, we see the same problems happening with nations collapsing from within because they ignore the type of guiding advice and wisdom available in our manufacturer's manual, the Holy Bible.

The four nations we are focusing upon, started well but will not end well if upright principles of leaders and citizens are not practiced, taught, and cherished. Greed, immorality, and other poor choices could bring our downfall because they are destructive practices. The signs are already evident.

This sounds like lecturing, and it is. A case can be made that the majority of us care about the well-being of our people and nations. Some wise person said words to the effect that ***real troubles come when good people do nothing.***

Good people can choose to practice positive principles in their community and help to repair what is deteriorating there and nationwide.

We, as individuals and nations, will enjoy life most by simply using the wisdom and advice of our Creator who cares so much and works so hard for each one of us every second of every day. We can define wisdom as the practical application of beneficial knowledge, coupled with consideration for others and common sense. Part of this is defending ourselves and our society from further destruction.

Understanding *temptation* is a huge part of wisdom. Temptations generally look attractive for a while, then they turn around and give us troubles that can ruin our life.

Having the wisdom to turn temptations off in their tracks, can protect us from all kinds of trouble.

I am lecturing here because I care so much for our nations and good citizens. I hate to see any in unnecessary pain and suffering when wisdom and good choices are so readily available.

Regarding the best choices in science, the original concept of following where best evidence leads is still the best path to choose.

We believe the science that we are currently calling "Atomic Biology" will lead to: an enjoyable appreciation of how much our Creator cares for us; a better understanding of our amazing bodies, minds, and gifts; happier attitudes; more wisdom for living; better food production; beneficial and remedial diets; better medicines; and better health.

In the four nations we have focused upon, we citizens are particularly blessed in many ways with plenty of food, shelter, clothing, comforts, health care, entertainment, etc. Our belief is that giving God "Thanks" at each mealtime and regularly counting our blessings is helpful in making our lives more enjoyable.

Hopefully, some principles and concepts in this book – and more so in the Bible – will help our governments to make more good choices resulting in at least some improved consequences nationwide.

Part II:

Teaching the "Why's" of God's Inclusion in Our Governments, and Why This Ties In with the Science of Atomic Biology

Chapter 7

Some Persons Dislike Their Concept of God

When you read the first few quotes at the beginning of Chapter 1, you may be a bit surprised to learn that Darwin was such a believer in our Creator.

Darwin's Theory has been misused in an effort to remove our Creator from public education. This ploy has been widely used in the last few decades to the detriment, not only to our students, but also to our governments and society in general.

The resulting moral decline and increase in depression and hopelessness have led to increasing drug use with all of its degenerative effects, overdose deaths, suicides, crime for addiction supplies, irresponsibility, higher costs to taxpayers for policing, emergency services, medical costs, family heartaches and tears.

Darwin's theory of evolution has been modified over the years, mainly by others who saw this as an opportunity to produce an image of the cause of life that is totally ungodly.

To the unbeliever, the written Word of the God of our nations – the Holy Bible – might seem like foolishness. There are aspects of this position that might seem reasonable, if you consider the following examples:

1. We cannot see God in person (except for the 33 years when Jesus was on Earth).

2. His voice is not often audible.

3. We might not believe in any miracles including those arranged by God.

4. It can seem foolish to believe in something by "faith" although belief in evolution now requires much more faith than belief in our Creator.

> Since belief in Evolution as the cause of life, knowing how brilliant and complex life is, requires immense Faith, this makes Evolution as much of a Religion as any other Belief requiring Faith.

Many people dislike their idea of God for various other reasons. Consider these examples:

1. They may have been offended or hurt by someone claiming to be a Godly person or attending a church. Most of us have heard stories about adults abusing children in some church-related setting or other. This is a truly disgusting action of a tempted pretender.

> In His "Word" – the Holy Bible – God says that it would be better for a person to have a millstone placed around his neck and be flung into the sea, than to harm one of His little ones. He says that He judges offenders after the end of their life and sometimes before.
>
> We have to understand that going to a church does not make anyone a Godly person any more than standing in a garage makes them a car.

2. Some people do not like their concept of God because He seems like an intruder who watches every move they make, including, perhaps, some they may not want watched.

3. They know some people who are church-goers who are imperfect (which includes all of us); they do bad things like anybody else, or worse, and don't stop.

4. Some people have not found the reasons to believe in God.

An important objective of this book is to show the reader just how much our Creator, God, cares for each one of us and how hard He works for us in providing for our lives every second of every day.

Here are a few thoughts that might become helpful for those who have trouble believing there is a Creator God who cares for us immensely, and who proves this by His enormous works in providing life, food, and beautiful things for us to enjoy.

Regarding point 1 above, it is true that we cannot see God. However, many things that exist cannot be seen. Here are just a few examples from hundreds:

i. We cannot see the wind, yet it is easy to see its effects on trees, flowers, clouds, and so on;

ii. We cannot see gravity, but we know it is there because if we drop something heavier than a helium balloon, it falls down;

iii. We cannot see the individual atoms and molecules we are built with but we know scientifically they are there;

iv. We cannot see warmth from the sun but we can feel it on a summer day;

v. We cannot see love but we instinctively feel it emotionally when it is close to us.

Regarding point 2, it is true that relatively few people have heard God's voice, however, there have been many reports of hearing audible words from God at significant times such as His audible warning at times of imminent danger while driving, boating, and so on.

Far more significant are the large numbers of people who experience answers to their prayers to God. No doubt, presidents and prime ministers would not pray privately and publicly if they believed it was a futile exercise.

In my studies about God and His Word – the Holy Bible – I have learned and experienced that there indeed are four separate types of answers to prayers, just as I was taught. All of these types of answers begin with "D".

i. First is the *Direct* answer, which comes while you are praying, or very shortly thereafter. For example, I have several times prayed to God for His help in finding my car keys, and other small items when I have become frustrated in searching for them, and before I have finished praying the request, I will be given the exact location.

ii. Next is the *Delayed* answer which comes after days, weeks, months, or even years of starting the prayer request. This type of answer often comes for those who have been praying for a spouse, the blessing of a baby child, for a special job, etc. For example, I prayed to God for several years for a job in His service, and finally it came, and that is the job of working on this book and project.

iii. Third is the *Different* answer whereby you pray for something in particular, but receive something else. Some young people might pray for a close relationship with a particular boy or girl, and that does not occur; instead, another individual –

often a better one – does become their special boyfriend or girlfriend.

iv. Fourth is the *Denied* answer which often happens because God knows what is best for us. He can deny the item we prayed for, but will often send something better for us.

In general, those who believe in God will probably say that they receive messages and advice from Him most often while reading or hearing His Word as written in the Holy Bible.

Regarding point 3, God's "'miracles" are well documented in many modern books, as well as in the Bible. Most people who have received them, probably give God the credit. I remember one of the first "miracle healings" that I played a small part in as a fairly new believer at age 55. Katherine, a young mother of three little children, had a very painful shoulder and could not move her right hand and arm away from her hip without severe pain. Her brother came to me and three others and asked if we would pray to God to heal her shoulder. After all Katherine had to be able to handle her three kids. So we each laid one of our hands on her back or shoulders and prayed for her healing for about two minutes. Then the other four people just left. So, as a newish believer, I asked, "Did that help any, Kathy?" Well, she lifted her arm right up straight over her head without any pain at all! "Praise the Lord!" I said. This was exciting for me as I had not seen – up close and personal – an instant healing like this before.

Regarding point 4, although believing in something by faith might not seem very solid or scientific, I have found that people generally have reasons for doing what they do and believing what they believe. Part of the objective of this primer is to provide solid reasons to believe and have faith in our Creator, God – scientific evidence here and now which can solidify our faith.

I agree with the message in the title of a book by Norman Geisler and Frank Turek, "*I Don't Have Enough FAITH to Be an ATHEIST.*" Knowing how highly complex, precisely built, and finely tuned virtually all necessities for life are, believing they could just happen with no intelligent input takes incredible faith.

Regarding point 5, we have to remember that God is not the church and people are not God. We all have an enemy within us who will tempt us to do bad or even evil deeds. This is Satan of whom you have probably heard, but might not understand.

Satan is bright, evil, powerful, and in competition with God for our attention. He is far smarter and stronger than we are on our own. He is very subtle and can make things that are bad for us, look good and attractive. For example, to many he will say things like, "Try a little of this crystal meth…you will feel fantastic!" And very shortly, many are trapped in one of the ugliest addictions in the world. And of course, the temptation to use fentanyl can be absolutely deadly. Here in the City of Vancouver (not even Greater Vancouver with 4 times the population) just last week, February 26th to March 5th, 2017, there were 174 emergency calls and 14 deaths due to drug overdoses, up 6 from the previous week.

Satan tries, on a regular basis, to lure each one of us into one of his traps. It seems that he delights in picking on people in churches in order to win them to his flock and make the Church, which is often considered a house of God, to look bad or even disgusting.

We just have to remember that God is not the church, and we cannot blame Him or His son Jesus, for what people do wrong.

Regarding point 6, God won't stop anyone from doing anything, and He does not report to anyone (although it seems some mothers get information on what their kids are doing wrong without seeing it). But because He has so much work to do within each of us, including processing our digesting foods, sustaining us, maintaining us, repairing, and replacing our worn-out cells with the new ones He has built for us, including brain cells, etc., He just has to be there. He might warn us not to do something through our conscience, but He gives us the freedom to choose what we do.

Regarding point 7, it is true that many church-goers choose to do bad things, and they do give the church a bad name. However, it is not God's will or teaching that directs their choices. More likely, it is Satan.

Just remember, God is not the church, and we should not judge Him by what people do. He works very hard to provide our life, along with wonderful things for us to enjoy, like beautiful birds, flowers, fish, animals, good friends, and pets. Our choice to do bad things is not His choice. He tries to teach us how to have a joyful life, to avoid bad choices, and to help others.

Regarding point 8, it is true that some people do not believe in the existence of God, and others do not want to believe in His existence for their own reasons. Hopefully, after reading this book, they will have several good scientific reasons to understand and appreciate His existence, especially with this new knowledge regarding all the caring work that He performs reliably for each one of us every second of every day.

Does God *deserve* our appreciation?

Chapter 8

A Fresh Introduction To The Scientific God Of Our Nations

As mentioned earlier, evidence is growing to indicate that most sciences are studies of the work our Creator is already doing.

Since the majority of the adult citizens in the four nations we are discussing, already believe in God, only more details are needed for them. But for those who have little knowledge of their Creator, we provide some basic information regarding the enormous amount of super-intelligent, caring work He performs for each one of us every day.

As one story goes, a little boy – who was probably speaking for many of us – said, "I want to see God with skin on!" – And another story representing the mindset of many of us, "I'm from Missouri…SHOW ME !"

This we endeavor to do…provide some ways to understand and measure God's immense care for each one of us.

Let us eliminate the gods we are not talking about here, i.e. the mythical Greek, Roman, Egyptian, or any other mythical god. We are referring to the Triune Creator God of the Holy Bible recognized and acknowledged by the governments of the United States, the United Kingdom, Australia, and Canada, as well as other national, state, provincial, county, and municipal governments.

This Bible is the 'Manufacturer's Manual for Operators' (us). As with other operator's manuals, its purpose is to supply advice for the user's maximum benefit.

God's guidance has always been available for the governments of these nations, and for the citizens who desire His counsel.

Details that, in the past, have not been clarified, are those showing the enormous amount of *careful physical work* He performs for everybody, including atheists and evolutionists, every second of every day.

Although we are providing some examples, so much more will be found through further research and investigation.
How do we know it is our Creator God who is doing all of this phenomenal work?

First, we analyze the super-intelligent, physical work *required* to build every morsel of our food, i.e. each vegetable, fruit, kernel of grain, etc., then we analyze the even-more-brilliant essential work required to assemble our cells using many of the atoms in our food, and then to place, fasten, and hook up each one of our cells to create, sustain, maintain, and repair *us*.

Each part of these various cells is made of the right numbers of the right atoms, and we know that atoms don't have legs, wings, fins, brains, or any other *internal* means to move themselves into a precise position in each cell. Therefore, a super-intelligently intelligent, dexterous, precise, speedy, and caring *external force* is required to do all this work.

The name given to this force by the English-speaking people and the governments of our four nations (and others) is "God".

We have investigated a portion of the work that we know requires *super-intelligent vision* to find in the soil, air, and water, all the right numbers of the right types of atoms necessary for the building blocks of every tiny root cell for our fruit and vegetables, every cell for the body of these food items, every cell for the leaves, skin, etc. Then we know that *super-intelligent dexterity* is required to sort, select, and grasp each of these right atoms; super-intelligent *precision and intelligence* are required to place each atom carefully into each cell-part the builder is creating for our food. Then, after we have eaten the food He has constructed for us, more *super-intelligent work* must be performed to find, sort, select, grasp, precisely place, and fasten each correct atom for each molecule for each different part of each different cell required for the construction, sustenance, growth, maintenance, and repair of our bodies, every second of every day.

We know by observation of growth of various foods that this brilliant work is going on constantly.

We know that we humans, with all our accumulated scientific intelligence and sophisticated equipment cannot make even one significant part of a cell, using pure elements and given any length of time.

We know, therefore, that to accomplish the enormous amount of brilliant works essential to build entire food cells and human cells, a super-intelligent and super-intelligent force is required. Building each cell-part using inanimate atoms is super-intelligent enough; however, without adding *the super-intelligent breath-of-life*, no cell or entity can live or function. When this is removed, the cell dies.

The production of our food, as one example, has been acknowledged for centuries, as special creation, and long before these details were understood, God was acknowledged as the brilliant, super-intelligent producer.

Governments have provided a special day of appreciation for the enormous work that God does for us all, and that is Thanksgiving or Thanksgiving Day.

To understand even a very small portion of the amazing *work* that has to be performed for us by this super-intelligent force, let's look at the following example.

Let us think for a few minutes of a desire to build a new, personal home where we could live. What work must be done in order for this to be started and completed?
We can relate this to building a body for a person, which, incidentally, is almost infinitely more complex than building a home.

First, you need a suitable piece of land for the home, and a mother's womb for a person. The land needs to be buildable and preferably have water, sewer, electricity, natural gas, television, and telephone services available. The mother's womb has to be reasonably healthy and capable of receiving the necessary atoms, molecules, and cells as building blocks for the construction of a baby.

For the home, a *plan* is required that shows whether there is or isn't a basement or a crawl space, whether there will be one, two, or three floor levels over the foundation; the exterior size of the building; the size of the foundation walls, height, and thickness, the level of the first floor above ground; what materials to use for the foundation and how they can be properly placed, where the materials can be acquired; and who is going to do the work for building the foundation. Of course, the materials for the foundation have to be designed, manufactured, and available from another location.

For building the body of a baby person, a *plan* is also required. Some of the details of the plan have to be brilliantly and carefully made and placed into each one of the father's millions of tiny sperm cells, and other cooperative plans have to be brilliantly and carefully made and placed into the mother's egg cells. The materials for building the sperm and egg cells have to be designed, manufactured, and available for delivery from another location away from the womb also.

For the home plan, the sizes of the rooms have to be designed: where the kitchen will be; where the bathroom(s) will be; and the living room, dining room, bedroom(s), closets, windows, doors, furnace, air conditioner(s), hot water heater, drains, ducts, lights, switches, plugs, chimney, stove, fridge, cabinets, plumbing, wiring, walls, roof supports, rafters, insulation, interior and exterior wall-coverings, roofing materials, stair locations, height of steps, etc.

It takes an intelligent designer to produce the plans for a home. To design a plan for a person requires *super-intelligent* intelligence, far beyond that of any human group. Not only does the structure have to be suitably designed for his or her overall functions, but each one of the approximately 100 trillion complex cells in an average adult, has to be designed to live, to function, and to reliably and consistently perform many specific living tasks for anywhere from a few days to many years before God builds and installs a new replacement for it.

Each person has to choose the good foods and beverages from which God can obtain the best building-block atoms to make his or her cells. An intelligent lifestyle is also important for healthy results.

The work involved in building a nice home requires considerable knowledge and several skills, but the construction of even a huge palace is *absolutely simple* in comparison to the knowledge, skills, dexterity, precision, care, and speed required in building each and every cell of a person.

Once a home is built and finished, it requires a certain amount of care and maintenance.

However, once a baby is built and delivered into the world, it requires enormous amounts of careful work, not just to feed, clothe, wash, change diapers, teach, and other external acts, but far, far more complex and demanding is the necessary internal work of building trillions more special cells in a carefully planned and critically ordered manner to grow the baby up to adult size, to manufacture the food required for the baby's growth and sustenance, to maintain all of the child's cells, and to repair or replace all those cells that constantly need sustaining, maintaining, repairing, or replacing.

One of the favorite parts we, at Reality R&D, like to look at is our red blood cells. We can tell that there is a mind-boggling amount of necessary work performed for us every second of every day just in replacing all of our roughly 20 trillion red blood cells about every 120 days. The number of right atoms that must be found in our eaten foods, then sorted, selected, counted, grasped, *precisely assembled*, and delivered into our bloodstream is *over 4,900 quadrillion of the right atoms every second of every day* for every adult. [1]

Of course, *this is less than half the intelligent physical work* required to manufacture and deliver our new replacement red blood cells, as, in the same second, a greater number of right atoms must be found in the soil, air, and water in gardens and fields, etc. to be biologically constructed into more foods for each future second's requirements for our replacement red blood cells.

A good question to ask is, "*Why* does this super-intelligent force we call 'God' work *so hard, so reliably, and so constantly* for each one of us?

We sincerely believe it must be because He loves us and cares immensely for each one of us.

Since He provides all of this necessary work, care, and much more to us reliably and free of charge for our whole lifetime, it is understandable why we can all rightfully say, *"In God We Can Trust"*.

In the glossary we defined God as: "the name given by English-speaking people and governments to the super-intelligent force and entity who creates and sustains all other living entities....."

Although we are referring to the four nations, there are many other peoples in countries where English is not the primary language, that acknowledge this same Creator God, but they have different names for Him. There are believers in God in China, India, Russia, France, Germany, South Korea, Israel, South Africa, Mexico, New Zealand, and many other nations. In fact, there are people in virtually all nations of the world who acknowledge God as the Creator of life and the universe – they just use non-English names for Him.

In this book you will see that we use a capital H for Him out of respect and appreciation for the enormous amount of faithful, reliable, and careful work that He performs for every person every day. In this book and in other textbooks to come, we will outline some of the main parts of His work for all of us.

This information is needed primarily because a deliberate exclusion of teachings about the God of our nations has occurred in most government-funded schools, colleges, and universities. This exclusion seems to be based mainly on the misconception that 'evolution' is more scientific than God.

As we have pointed out before, nothing could be further from the truth.

However, in fairness, many of the key, scientific details of how God performs His awesome works for us were not previously understood or available as they are now.

Charles Darwin released the first edition of his book, "*On the Origin of Species by Means of Natural Selection, or The Preservation of Favoured Races in the Struggle for Life*" in 1859. It made reference to his theory of evolution. He theorized that perhaps all life had originated from a unique happening in an ancient pond where the first living organism was mysteriously formed without any intelligent help.

This organism or cell would have needed the complex ability, power, and intelligence to live; to find nourishment; to digest and process this nourishment to sustain its own life; to eliminate waste so as not to build up toxins; to produce offspring; to remember its good qualities; to pass these on to new cells with improvements; to produce more cells; and over billions of years, to change into fish, animals, birds, monkeys, humans, etc. All this with no intelligent help.

Although this was Mr. Darwin's idea of a possibility, he also mentioned the "Creator" several times. For example, in his book on '*Origins*', Darwin says of eyesight: *"[M]ay we not believe that a living optical instrument might thus be formed as superior to one of glass, as the Works of the Creator are to those of man?"* [2]

Obviously, Darwin had great respect for the Creator and His superiority above man's abilities.

He also outlined several ways that his theory of evolution could "absolutely break down."

We want to show a number of the reasons why it is absolutely impossible for any living cell, organ, or living entity to be designed, constructed, sustained, maintained, repaired, and replaced by anything but the super-intelligent physical works of a super-intelligent force.

In Chapter 5 we provide seven basic principles of life indicating the need for a highly intelligent, hard-working, immensely reliable, trustworthy, and caring super-intelligent force to design, construct, and do all the brilliant work essential for all living entities, and especially us human beings.

Chapter 5 also provides a list of eighteen e*ssential intelligent physical works* required to build and maintain each living entity *which Evolution cannot do, as, by definition, it has no intelligence.* We call this "The Solid Evidence" for the super-intelligent force we call "God."

At the time of the founding of the four nations we are addressing, not all of this science was known or available in detail for our founding fathers. However, they understood and appreciated the need for super-intelligent external help with growing crops of grains, vegetables, and fruit for food, and the production of cattle for milk and meat, fish and fowl to eat; they knew it was reliably and faithfully *created* for the benefit of humankind.

They called this super-intelligent external creator and helper God, just as the majority of citizens do today in these four nations.

This is the same God of the Holy Bible who has been known and respected for at least the past forty-two hundred years. God's word, the "Holy Bible," has been the annual best-selling book in the world for centuries and has been translated into hundreds of languages and dialects.

In later chapters, which outline the role of God in the governments of our nations, we will see that many of the founding fathers took time to study the wise and beneficial advice in God's inspiring "Word," the Holy Bible.

This point is made, not to show prejudice or favoritism, but as part of the facts that help to reveal the sources of the success that these four nations have enjoyed. These successes have attracted millions of immigrants from countries with other beliefs.

Today, many of our leaders pray to God for help in times of national and international distress and for His blessings for their nation and people.

Do not be fooled by those who say there is no God. – There obviously *is* and we can see phenomenal amounts of real evidence of His works in our own lives and in all life around us, when we understand at least some of the details.

He shows how much He cares for us by working very hard, consistently, reliably, faithfully, and brilliantly for each one of us every second of every day.

When we consider all this phenomenal work that God performs for us, it is only good manners for us to thank Him. Thanksgiving Day was appropriated especially for that purpose, but it is just good manners to say "thank you, God" often, for meals, healing, advice, prayer answers, and so on.

We may never know of *all* the work He performs for us.

Many people seem to believe that throwing away our moral compass is a great benefit for our society, or at least for them individually. However, the statistics showing the skyrocketing increase in drug abuse, alcoholism, teen pregnancies, depression, suicides, drug addiction, overdose deaths, and all the family break-ups, trauma, and tears since the turning point in 1963, prove otherwise.

In 1963, Madeline Murray (O'Hare) with some of her atheist friends, managed to convince the US Supreme Court members to officially remove prayer from public school classrooms. The concept spread to other nations. Since then, attempts are being made to remove all aspects of the God of our nations from public life.

History and the current statistics show what a destructive idea this is. Reversing this trend will take wisdom, care, and cooperation from concerned citizens.

You can help.

Chapter 9
God in the Government of the U.S.A.

Stephen B. Goodwin/Shutterstock.com

CAPITOL HILL, U.S.A.

This chapter and the science chapters show many reasons why students in the USA have the inalienable right to be taught about the God of their nation. He is a significant part of their government and can provide wisdom, hope, encouragement, grace, understanding, and more accurate knowledge of life.

Examples of where God has been, and is now, involved in the government of this nation:

- the Declaration of Independence;
- the Articles of Confederation;
- the Constitution;
- the Pledge of Allegiance;
- the Presidential Inaugural Addresses;
- the currency;
- the national anthem and songs;
- the courts and justice systems;
- public buildings and places;
- war memorials;
- leaders' prayers;
- public holidays;
- majority are believers.

We will list some examples of the times that prayer is used at government functions and meetings of various levels of leadership. All of these examples are shown so that citizens can see some of the areas of government where God is being called upon to help improve the nation with His wisdom, power, capability, advice, and blessings.

As there has been much debate regarding the meaning of the concept of the "separation of church and state", we must clarify that our Creator God is *not* "the church". These days "the church" can be a group of people or a building for Satanists, Scientologists, Christians, Muslims, Mormons, Jehovah's Witnesses, New Agers, cults, etc. God is definitely NOT "the church".

God is, however, a highly-regarded part of the Government of the United States as is witnessed by the reference to Him in the examples detailed in this chapter.

This high regard for God has been present in North America since before there were any formal States to be united.

The following examples of God's recognition show a few of the reasons for His importance to the nation.

Note: Author's emphasis has been added with underlining.

1. The Declaration of Independence: (Excerpts)

"In Congress, July 4, 1776. The unanimous Declaration of the thirteen United States of America:

When in the Course of human events, it becomes necessary for one people to dissolve the political bands which have connected them with another, and to assume among the powers of the earth, the separate and equal station to which the Laws of Nature and of Nature's God entitle them, a decent respect to the opinions of mankind requires they should declare the causes which impel them to the separation.

We hold these truths to be self-evident, that all men are created equal, that they are endowed by their Creator with certain unalienable Rights, that among these are Life, Liberty, and the pursuit of Happiness. " ...

(2nd to last paragraph) "*We, therefore, the Representatives of the United States of America, in General Congress, Assembled, appealing to the Supreme Judge of the world for the rectitude of our intentions, do, in the Name, and by the Authority of the good People of these Colonies, solemnly declare, That these United*

Colonies are, and of Right ought to be, Free and...Independent States;... [1]

2. The Articles of Confederation: (Excerpts)

The thirteen Articles were officially adopted on November 15th, 1777 by representatives of the original thirteen states.

The second paragraph of Article XIII reads: *"And whereas it hath pleased the Governor of the World to incline the hearts of the legislatures we respectively represent in congress, to approve of, and to authorize us to ratify the said articles of confederation and perpetual union."* [2]

3. The United States Constitution:(Excerpts)

This document signed September 17, 1787, established the Government of the United States.

As a major goal of many of the early immigrants to the land was to have freedom of religion, the U.S. Constitution helps to ensure that freedom by including the following statement in Clause 3 of Article 6: " *...no religious test shall ever be required as a qualification to any office or public trust under the United States."*[3]

Believing in God is not a legal requirement to run for high office in government. There is freedom to choose what you want to believe in, what or whom you want to have faith in, or what or whom you want to worship, if anything or anyone. It only makes good sense to choose the one that does the most for you.

4. The Pledge of Allegiance: (Excerpts)

The original version of the Pledge of Allegiance was published on September 8, 1892. It was refined periodically over the years until the present version was adopted on Flag Day, June 14th, 1954: *"I pledge allegiance to the Flag of the United States of America, and to the Republic for which it stands, one nation under God, indivisible, with liberty and justice for all."* [4]

5. First Presidential Inaugural Addresses:

(Excerpts are courtesy of the Lillian Goldman Law Library's Avalon Project, Yale University Law School) [5]

Note: Every President of the USA has included a request to God for His blessing on the nation, in their inaugural addresses as well as many times during their terms as President.

Only a few excerpts are shown herein:

President George Washington's First Inaugural Address, April 30, 1789: (Excerpts)

> *"Having thus imparted to you my sentiments as they have been awakened by the occasion which brings us together, I shall take my present leave; but not without resorting once more to the benign Parent of the Human Race in humble supplication that, since He has been pleased to favor the American people with opportunities for deliberating in perfect tranquility, and dispositions*

for deciding with unparalleled unanimity on a form of government for the security of their union and the advancement of their happiness, so His divine blessing may be equally conspicuous in the enlarged views, the template consultations, and the wise measures on which the success of this Government must depend."

President Thomas Jefferson's First Inaugural Address, March 4, 1801: (Excerpts)

"Relying, then, on the patronage of your good will, I advance with obedience to the work, ready to retire from it whenever you become sensible how much better choice it is in your power to make. And may that Infinite Power which rules the destinies of the universe lead our councils to what is best, and give them a favorable issue for your peace and prosperity."

President Abraham Lincoln's First Inaugural Address, March 4, 1861: (Excerpts)

"Why should there not be a patient confidence in the ultimate justice of the people? Is there any better or equal hope in the world? In our present differences, is either party without faith of being in the right? If the Almighty Ruler of Nations, with His eternal truth and justice, be on your side of the North, or on yours of the South, that truth and that

justice will surely prevail by the judgment of this great tribunal of the American people."

".... Intelligence, patriotism, Christianity, and a firm reliance on Him who has never yet forsaken this favored land are still competent to adjust in the best way all our present difficulty."

"You have no oath registered in heaven to destroy the Government, while I shall have the most solemn one to "preserve, protect, and defend it."

President Lincoln's Gettysburg Address, November 19, 1863: (Excerpts)

"Four score and seven years ago our fathers brought forth on this continent a new nation, conceived in liberty and dedicated to the proposition that all men are created equal. Now we are engaged in a great civil war, testing whether that nation or any nation so conceived and so dedicated can long endure. We are met on a great battlefield of that war. We have come to dedicate a portion of that field as a final resting-place for those who here gave their lives that that nation might live. It is altogether fitting and proper that we should do this. But in a larger sense, we cannot dedicate, we cannot consecrate, we cannot hallow this ground. The brave men, living and dead who struggled here have consecrated it far above our poor power to add or detract. The world will little note nor long remember what we say here, but it can never forget what they did here. It is for us the

living rather to be dedicated here to the unfinished work which they who fought here have thus far so nobly advanced. It is rather for us to be here dedicated to the great task remaining before us – that from these honored dead we take increased devotion to that cause for which they gave the last full measure of devotion – that we here highly resolve that these dead shall not have died in vain, that this nation under God shall have a new birth of freedom, and that government of the people, by the people, for the people shall not perish from the earth."

President Franklin D. Roosevelt's Fourth Inaugural Address, January 20, 1945: (Excerpts)

"As I stand here today, having taken the solemn oath of office in the presence of my fellow countrymen—in the presence of our God—I know that it is America's purpose that we shall not fail."

"In the days and in the years that are to come we shall work for a just and honorable peace, a durable peace, as today we work and fight for total victory in war."

"The Almighty God has blessed our land in many ways. He has given our people stout hearts and strong arms with which to strike mighty blows for freedom and truth. He has given to our country a faith which has become the hope of all peoples in an anguished world."

"So we pray to Him now for the vision to see our way clearly—to see the way that leads to a better life for ourselves and for all our fellow men—to the achievement of His will to peace on earth."

President Harry S. Truman's Inaugural Address, January 20, 1949: (Excerpts)

"The American people stand firm in the faith which has inspired this Nation from the beginning. We believe that all men have a right to equal justice under law and equal opportunity to share in the common good. We believe that all men have the right to freedom of thought and expression. We believe that all men are created equal because they are created in the image of God."

"From this faith we will not be moved."

President Dwight D. Eisenhower's First Inaugural Address, January 20, 1953: (Excerpts)

"My friends, before I begin the expression of those thoughts that I deem appropriate to this moment, would you permit me the privilege of uttering a little private prayer of my own. And I ask that you bow your heads"

"Almighty God, as we stand here at this moment my future associates in the executive branch of government join me in beseeching that Thou will make full and complete our dedication to the

service of the people in this throng, and their fellow citizens everywhere."

"Give us, we pray, the power to discern clearly right from wrong, and allow all our words and actions to be governed thereby, and by the laws of this land. Especially we pray that our concern shall be for all the people regardless of station, race, or calling."

"May cooperation be permitted and be the mutual aim of those who, under the concepts of our Constitution, hold to differing political faiths; so that all may work for the good of our beloved country and Thy glory. Amen."

"My fellow citizens:

The world and we have passed the midway point of a century of continuing challenge. We sense with all our faculties that forces of good and evil are massed and armed and opposed as rarely before in history.

This fact defines the meaning of this day. We are summoned by this honored and historic ceremony to witness more than the act of one citizen swearing his oath of service, <u>in the presence of God</u>. We are called as a people to give testimony in the sight of the world to our faith that the future shall belong to the free.

In the swift rush of great events, we find ourselves groping to know the full sense and meaning of these times in which we live. In our quest of understanding, we beseech God's guidance.

It is because we, all of us, hold to these principles that the political changes accomplished this day do not imply turbulence, upheaval or disorder. Rather this change expresses a purpose of strengthening our dedication and devotion to the precepts of our founding documents, a conscious renewal of faith in our country and in the watchfulness of a Divine Providence.

The enemies of this faith know no god but force, no devotion but its use. They tutor men in treason. They feed upon the hunger of others. Whatever defies them, they torture, especially the truth.

Here, then, is joined no argument between slightly differing philosophies. This conflict strikes directly at the faith of our fathers and the lives of our sons. No principle or treasure that we hold, from the spiritual knowledge of our free schools and churches to the creative magic of free labor and capital, nothing lies safely beyond the reach of this struggle.

Freedom is pitted against slavery; lightness against the dark.

This is the hope that beckons us onward in this century of trial. This is the work that awaits us all, to be done with bravery, with charity, and with prayer to Almighty God."

President John F. Kennedy's Inaugural Address, January 20, 1961: (Excerpts)

"Vice President Johnson, Mr. Speaker, Mr. Chief Justice, President Eisenhower, Vice President Nixon, President Truman, reverend clergy, fellow citizens, we observe today not a victory of party, but a celebration of freedom – symbolizing an end, as well as a beginning – signifying renewal, as well as change. For I have sworn before you and Almighty God the same solemn oath our forebears prescribed nearly a century and three quarters ago."

"The world is very different now. For man holds in his mortal hands the power to abolish all forms of human poverty and all forms of human life. And yet the same revolutionary beliefs for which our forebears fought are still at issue around the globe – the belief that the rights of man come not from the generosity of the state, but from the hand of God. ..."

"Let both sides unite to heed in all corners of the earth the command of Isaiah – to "undo the heavy burdens ... and to let the oppressed go free."

"Finally, whether you are citizens of America or citizens of the world, ask of us the same high standards of strength and sacrifice which we ask of you. With a good conscience our only sure reward, with history the final judge of our deeds, let us go forth to lead the land we love, asking His blessing

and His help, but knowing that here on earth God's work must truly be our own."

President Lyndon B. Johnson's Inaugural Address, January 20, 1965: (Excerpts)

"My fellow countrymen, on this occasion, the oath I have taken before you and before God is not mine alone, but ours together. We are one nation and one people. Our fate as a nation and our future as a people rest not upon one citizen, but upon all citizens. ..."

"Our destiny in the midst of change will rest on the unchanged character of our people, and on their faith."

"We have discovered that every child who learns, every man who finds work, every sick body that is made whole – like a candle added to an altar – brightens the hope of all the faithful."

"Let us now join reason to faith and action to experience, to transform our unity of interest into a unity of purpose. For the hour and the day and the time are here to achieve progress without strife, to achieve change without hatred—not without difference of opinion, but without the deep and abiding divisions which scar the union for generations."

"THE AMERICAN BELIEF:

Under this covenant of justice, liberty, and union we have become a nation – prosperous, great, and mighty. And we have kept our freedom. But we have no promise from God that our greatness will endure. We have been allowed by Him to seek greatness with the sweat of our hands and the strength of our spirit."

"Our enemies have always made the same mistake. In my lifetime—in depression and in war—they have awaited our defeat. Each time, from the secret places of the American heart, came forth the faith they could not see or that they could not even imagine. It brought us victory. And it will again."

President Jimmy Carter's Inaugural Address, January 20, 1977: (Excerpts)

"Here before me is the Bible used in the inauguration of our first President, in 1789, and I have just taken the oath of office on the Bible my mother gave me a few years ago, opened to a timeless admonition from the ancient prophet Micah:

'He hath showed thee, O man, what is good; and what doth the Lord require of thee, but to do justly, and to love mercy, and to walk humbly with thy God.' (Micah 6: 8)"

"Ours was the first society openly to define itself in terms of both spirituality and of human liberty. It is that unique self-definition which has given us an

exceptional appeal, but it also imposes on us a special obligation, to take on those moral duties which, when assumed, seem invariably to be in our own best interests."

"Let us learn together and laugh together and work together and pray together, confident that in the end we will triumph together in the right."

"Within us, the people of the United States, there is evident a serious and purposeful rekindling of confidence. And I join in the hope that when my time as your President has ended, people might say this about our Nation:

that we had remembered the words of Micah and renewed our search for humility, mercy, and justice;
that we had torn down the barriers that separated those of different race and region and religion, and where there had been mistrust, built unity, with a respect for diversity;
that we had found productive work for those able to perform it;
that we had strengthened the American family, which is the basis of our society;
that we had ensured respect for the law, and equal treatment under the law, for the weak and the powerful, for the rich and the poor;
and that we had enabled our people to be proud of their own Government once again."

President Ronald Reagan's First Inaugural Address, January 20, 1981: (Excerpts)

"We have every right to dream heroic dreams. Those who say that we are in a time when there are no heroes just don't know where to look. You can see heroes every day going in and out of factory gates. Others, a handful in number, produce enough food to feed all of us and then the world beyond. You meet heroes across a counter—and they are on both sides of that counter. There are entrepreneurs with faith in themselves and faith in an idea; who create new jobs, new wealth and opportunity. They are individuals and families whose taxes support the Government and whose voluntary gifts support church, charity, culture, art, and education. Their patriotism is quiet but deep. Their values sustain our national life.

I have used the words "they" and "their" in speaking of these heroes. I could say "you" and "your" because I am addressing the heroes of whom I speak—you, the citizens of this blessed land. Your dreams, your hopes, your goals are going to be the dreams, the hopes, and the goals of this administration, so help me God.

Above all, we must realize that no arsenal, or no weapon in the arsenals of the world, is so formidable as the will and moral courage of free men and women. It is a weapon our adversaries in today's world do not have. It is a weapon that we as Americans do have. Let that be understood by those who practice terrorism and prey upon their neighbors.

I am told that tens of thousands of prayer meetings are being held on this day, and for that I am deeply grateful. We are a nation under God, and I believe God intended for us to be free. It would be fitting and good, I think, if on each Inauguration Day in future years it should be declared a day of prayer.

Each one of those markers (in Arlington National Cemetery) *is a monument to the kinds of hero I spoke of earlier. Their lives ended in places called Belleau Wood, The Argonne, Omaha Beach, Salerno and halfway around the world on Guadalcanal, Tarawa, Pork Chop Hill, the Chosin Reservoir, and in a hundred rice paddies and jungles of a place called Vietnam.*

Under one such marker lies a young man – Martin Treptow – who left his job in a small-town barber shop in 1917 to go to France with the famed Rainbow Division. There, on the western front, he was killed trying to carry a message between battalions under heavy artillery fire.

We are told that on his body was found a diary. On the flyleaf under the heading, "My Pledge," he had written these words: "America must win this war. Therefore, I will work, I will save, I will sacrifice, I will endure, I will fight cheerfully and do my utmost, as if the issue of the whole struggle depended on me alone."

The crisis we are facing today does not require of us the kind of sacrifice that Martin Treptow and so many thousands of others were called upon to make. It does require, however, our best effort, and our willingness to believe in ourselves and to believe in our capacity to perform great deeds; to believe that together, with God's help, we can and will resolve the problems which now confront us.

And, after all, why shouldn't we believe that? We are Americans. God bless you, and thank you."

President George H. Bush's Inaugural Address, January 20, 1989: (Excerpts)

" *I have just repeated word for word the oath taken by George Washington 200 years ago, and the Bible on which I placed my hand is the Bible on which he placed his. It is right that the memory of Washington be with us today, not only because this is our Bicentennial Inauguration, but because Washington remains the Father of our Country. And he would, I think, be gladdened by this day; for today is the concrete expression of a stunning fact: our continuity these 200 years since our government began.*

We meet on democracy's front porch, a good place to talk as neighbors and as friends. For this is a day when our nation is made whole, when our differences, for a moment, are suspended.

And my first act as President is a prayer. I ask you to bow your heads:

Heavenly Father, we bow our heads and thank You for Your love. Accept our thanks for the peace that yields this day and the shared faith that makes its continuance likely. Make us strong to do Your work, willing to heed and hear Your will, and write on our hearts these words: 'Use power to help people.' For we are given power not to advance our own purposes, nor to make a great show in the world, nor a name. There is but one just use of power, and it is to serve people. Help us to remember it, Lord. Amen.

No President, no government, can teach us to remember what is best in what we are. But if the man you have chosen to lead this government can help make a difference; if he can celebrate the quieter, deeper successes that are made not of gold and silk, but of better hearts and finer souls; if he can do these things, then he must.

America is never wholly herself unless she is engaged in high moral principle.

There are few clear areas in which we as a society must rise up united and express our intolerance. The most obvious now is drugs. And when that first cocaine was smuggled in on a ship, it may as well have been a deadly bacteria, so much has it hurt the body, the soul of our country. And there is much to be done and to be said, but take my word for it: This scourge will stop.

And so, there is much to do; and tomorrow the work begins. I do not mistrust the future; I do not fear what is ahead. For our problems are large, but our heart is larger. Our challenges are great, but our will is greater. And if our flaws are endless, God's love is truly boundless.

Thank you. God bless you and God bless the United States of America."

President William Clinton's First Inaugural Address, January 20, 1993: (Excerpts)

"And so, my fellow Americans, at the edge of the 21st century, let us begin with energy and hope, with faith and discipline, and let us work until our work is done. The scripture says, 'And let us not be weary in well-doing, for in due season, we shall reap, if we faint not.'"

"From this joyful mountaintop of celebration, we hear a call to service in the valley. We have heard the trumpets. We have changed the guard. And now, each in our way, and with God's help, we must answer the call.

Thank you, and God bless you all."

President George W. Bush's First Inaugural Address, January 20, 2001: (Excerpts)

"America, at its best, is a place where personal responsibility is valued and expected. Encouraging

responsibility is not a search for scapegoats, it is a call to conscience. Though it requires sacrifice, it brings a deeper fulfillment. We find the fullness of life not only in options, but in commitments. We find that children and community are the commitments that set us free. Our public interest depends on private character, on civic duty and family bonds and basic fairness, on uncounted, unhonored acts of decency which give direction to our freedom. Sometimes in life we are called to do great things. But as a saint of our times has said, every day we are called to do small things with great love. The most important tasks of a democracy are done by everyone. I will live and lead by these principles, "to advance my convictions with civility, to pursue the public interest with courage, to speak for greater justice and compassion, to call for responsibility and try to live it as well." In all of these ways, I will bring the values of our history to the care of our times.

Americans are generous and strong and decent, not because we believe in ourselves, but because we hold beliefs beyond ourselves. When this spirit of citizenship is missing, no government program can replace it. When this spirit is present, no wrong can stand against it.

We are not this story's author, who fills time and eternity with His purpose. Yet His purpose is achieved in our duty, and our duty is fulfilled in service to one another. Never tiring, never yielding, never finishing, we renew that purpose today; to make our country more just and generous; to affirm the dignity of our lives and every life.

God bless you all, and God bless America."

President Barack Obama's First Inaugural Address, January 20, 2009: (Excerpts)

"We remain a young nation, but in the words of Scripture, the time has come to set aside childish things. The time has come to reaffirm our enduring spirit; to choose our better history; to carry forward that precious gift, that noble idea, passed on from generation to generation: the God-given promise that all are equal, all are free, and all deserve a chance to pursue their full measure of happiness."

"In the face of our common dangers, in this winter of our hardship, let us remember these timeless words. With hope and virtue, let us brave once more the icy currents, and endure what storms may come. Let it be said by our children's children that when we were tested we refused to let this journey end, that we did not turn back, nor did we falter; and with eyes fixed on the horizon and God's grace upon us, we carried forth that great gift of freedom and delivered it safely to future generations."

"Thank you. God bless you. And God bless the United States of America."

President Donald Trump's First Inaugural Address, January 20, 2017: (Excerpts)

"We, the citizens of America, are now joined in a great national effort to rebuild our country and to restore its promise for all of our people.

"We will face challenges. We will confront hardships. But we will get the job done.

"For many decades, we've enriched foreign industry at the expense of American industry; subsidized the armies of other countries while allowing for the very sad depletion of our military; we've defended other nation's borders while refusing to defend our own; and spent trillions of dollars overseas while America's infrastructure has fallen into disrepair and decay.

"One by one, the factories shuttered and left our shores, with not even a thought about the millions upon millions of American workers left behind.

"But that is the past. And now we are looking only to the future. We, assembled here today, are issuing a new decree to be heard in every city, in every foreign capital, and in every hall of power.

"From this day forward, a new vision will govern our land.

"From this moment on, it's going to be America first. And whether a child is born in the urban sprawl of Detroit or the wind-swept plains of Nebraska, they look up at the same sky, they fill their hearts with the same dreams, and they are infused with the same breath of life by the same almighty Creator.

"And yes, together, we will make America great again. Thank you. God bless you. And God bless America.

Note: All presidential excerpts (except President Trump's, which came during this writing time) are courtesy of the Lillian Goldman Law Library's Avalon Project, Yale Law School. [5]

A major challenge for the Presidents, of course, is to ensure that they and their government body live up to their stated intentions. They are in fact, role models to many citizens, especially students and young adults. Their morality and short-comings are witnessed and often copied. *"If the president can do it, so can I."*

Therefore it is crucial for the benefit of the nation that the leader be of high moral character.

6. The Currency:
From the U.S. Department of the Treasury

"The Congress passed the Act of April 22, 1864. This legislation changed the composition of the one-cent coin and authorized the minting of the two-cent coin. The Mint Director was directed to develop the designs for these coins for final approval of the Secretary. IN GOD WE TRUST first appeared on the 1864 two-cent coin. ....

"Another Act of Congress passed on March 3, 1865. It allowed the Mint Director, with the Secretary's approval, to place the motto on all gold and silver coins that "shall admit the inscription thereon." Under the Act, the motto was placed on the gold double-eagle coin, the gold eagle coin, and the gold half-eagle coin. It was also placed on the silver dollar coin, the half-dollar coin and the quarter-dollar coin, and on the nickel three-cent coin beginning in

1866. Later, Congress passed the Coinage Act of February 12, 1873. It also said that the Secretary "may cause the motto IN GOD WE TRUST to be inscribed on such coins as shall admit of such motto." .....

"A law passed by the 84th Congress (P.L. 84-140) and approved by the President on July 30, 1956, the President approved a Joint Resolution of the 84th Congress, declaring IN GOD WE TRUST the national motto of the United States. IN GOD WE TRUST was first used on paper money in 1957, when it appeared on the one-dollar silver certificate. The first paper currency bearing the motto entered circulation on October 1, 1957. *The Bureau of Engraving and Printing (BEP) was converting to the dry intaglio printing process. During this conversion, it gradually included IN GOD WE TRUST in the back design of all classes and denominations of currency.*

From the U.S. Department of the Treasury. [6]

7. The U.S. National Anthem:

"President Woodrow Wilson ordered the playing of *The Star-Spangled Banner*, at military and naval occasions in 1916, but it was not designated the national anthem by an Act of Congress until 1931.

"The words were written in 1814 by Francis Scott Key, who had been inspired by the sight of the American flag still flying over Fort McHenry after a night of heavy British bombardment. The text was immediately set to a popular melody of the time, *To Anacreon in Heaven.*

"The National Anthem consists of four verses, although on most occasions, only the first verse is sung:

Oh, say can you see by the dawn's early light
What so proudly we hailed at the twilight's last gleaming?
Whose broad stripes and bright stars thru the perilous fight,
O'er the ramparts we watched were so gallantly streaming?
And the rockets' red glare, the bombs bursting in air,
Gave proof through the night that our flag was still there.
Oh, say does that star-spangled banner yet wave
O'er the land of the free and the home of the brave?

On the shore, dimly seen through the mists of the deep,
Where the foe's haughty host in dread silence reposes,
What is that which the breeze, o'er the towering steep,
As it fitfully blows, half conceals, half discloses?
Now it catches the gleam of the morning's first beam,
In full glory reflected now shines in the stream:
'Tis the star-spangled banner! Oh long may it wave
O'er the land of the free and the home of the brave.

And where is that band who so vauntingly swore
That the havoc of war and the battle's confusion,
A home and a country should leave us no more!
Their blood has washed out of their foul footsteps' pollution.
No refuge could save the hireling and slave'
From the terror of flight and the gloom of the grave:
And the star-spangled banner in triumph doth wave
O'er the land of the free and the home of the brave.

Oh! thus be it ever, when freemen shall stand
Between their loved home and the war's desolation!
Blest with victory and peace, may the heav'n rescued land
Praise the Power that hath made and preserved us a nation.

Then conquer we must, when our cause it is just,
And this be our motto: 'In God is our trust.'
And the star-spangled banner in triumph shall wave
O'er the land of the free and the home of the brave." [7]

Also worth considering are the lyrics to *God Bless America,* written by Irving Berlin in 1918 and revised by him twenty years later. Like *The Star-Spangled Banner,* this tune is sung at many government functions:

God bless America,
Land that I love,
Stand beside her and guide her
Thru the night with a light from above;
From the mountains, to the prairies,
To the oceans white with foam,
God bless America,
My home, sweet home,
God bless America,
My home, sweet home.

8. U.S. Supreme Court and Justice Systems:

"When the Supreme Court opens, the Justices enter and all persons attending stand. They also stand as the Marshal of the Court chants, '*The Honorable Chief Justice and the Associate Justices of the Supreme Court of the United States. Oyez! Oyez! Oyez! All persons having business before the Honorable, the Supreme Court of the United States, are admonished to draw near and give their attention, for the Court is now sitting. God save the United States and this Honorable Court!"* [8]

The U.S. Supreme Court hears various arguments, and issues some one hundred and fifty annual major interpretations of the U.S. Constitution.

9. The Constitutional Oath of Office (5 USC § 3331):

Justices of the Supreme Court of the United States are required to take two oaths before they may execute the duties of their appointed office.

i. **"The Constitutional Oath**

All federal officials must take an oath in support of the Constitution. The Constitution does not provide the wording for this oath, leaving that to the determination of Congress. From 1789 until 1861, this oath was, *"I do solemnly swear (or affirm) that I will support the Constitution of the United States."* During the 1860s, this oath was altered several times before Congress settled on the text used today, which is set out at 5 U. S. C. § 3331. This oath is now taken by all federal employees, other than the President:

"I, _________, do solemnly swear (or affirm) that I will support and defend the Constitution of the United States against all enemies, foreign and domestic; that I will bear true faith and allegiance to the same; that I take this obligation freely, without any mental reservation or purpose of evasion; and that I will well and faithfully discharge the duties of the office on which I am about to enter. So help me God." [9]

ii. **The Judicial Oath:**

The origin of the second oath is found in the Judiciary Act of 1789, which reads *"the justices of the Supreme Court, and the district judges, before they proceed to execute the duties of their*

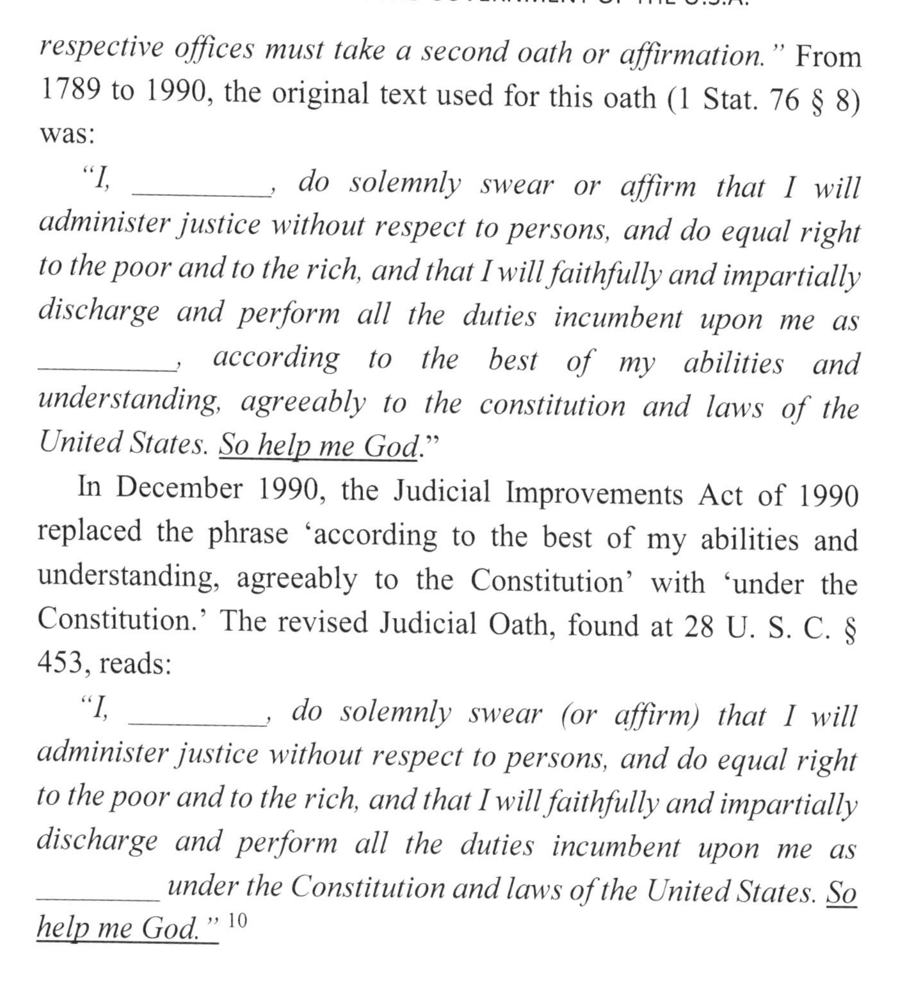

respective offices must take a second oath or affirmation." From 1789 to 1990, the original text used for this oath (1 Stat. 76 § 8) was:

"I, _________, do solemnly swear or affirm that I will administer justice without respect to persons, and do equal right to the poor and to the rich, and that I will faithfully and impartially discharge and perform all the duties incumbent upon me as _________, according to the best of my abilities and understanding, agreeably to the constitution and laws of the United States. So help me God."

In December 1990, the Judicial Improvements Act of 1990 replaced the phrase 'according to the best of my abilities and understanding, agreeably to the Constitution' with 'under the Constitution.' The revised Judicial Oath, found at 28 U. S. C. § 453, reads:

"I, _________, do solemnly swear (or affirm) that I will administer justice without respect to persons, and do equal right to the poor and to the rich, and that I will faithfully and impartially discharge and perform all the duties incumbent upon me as _________ under the Constitution and laws of the United States. So help me God." [10]

10. Public Buildings and Places:

The following are a few of the many places where God is referred to within, and on government buildings in Washington, D.C. For example, the Capitol Building bears many major paintings and inscriptions regarding our Creator God.

Permission by Architect of the Capitol

i. **"Embarkation of the Pilgrims"** – *Painting by Robert W. Weir (1803 – 1889)*

This 12-foot-high by 18-foot-wide painting in the Rotunda of the Capitol Building portrays the pilgrims in 1620 on their ship *Speedwell* just before they left Holland. The words *"God with us"* were written on the sail. (You have to look closely to see this in the upper-left corner). The painting shows the pilgrims with their Bible open, in prayer for success and safety in their great adventure. Their goal was to find a land where they could enjoy religious freedom.

Because of problems with the *'Speedwell,* they had to change ships in England. They booked passage on the *Mayflower* which was heading for America. This ship brought them to their new land where they established the Plymouth Colony, now known as Massachusetts.

Permission by Architect of the Capitol

ii. **"Declaration of Independence, July 4, 1776"** – *Painting by John Trumbull (1756 – 1843)*

This is another 12 foot by 18 foot painting in the Rotunda of the Capitol Building. It portrays the time when the first draft of the Declaration of Independence was presented to the members of the Second Continental Congress on June 28, 1776. The document's principles remain foundational for the nation today. On July 4, 1776, the delegates from the colonies signed *The Declaration of Independence* which marked a major milestone in the development of the United States of America.

In the painting's central group, Thomas Jefferson, the principal author of the Declaration, is shown presenting the document to John Hancock, the president of the Continental Congress. Near him are other members of the committee that created the draft: John Adams, Roger Sherman, Robert Livingston, and Benjamin

Franklin. The location is in what is now called Independence Hall, in Philadelphia.

iii. **Capitol Building inscriptions:**

In the Cox Corridor:

"America! God shed his grace on Thee, and crown thy good with brotherhood from sea to shining sea!"

—Katharine Lee Bates

In the House Chamber:

"In God We Trust"

In the Prayer Room:

"Annuit Coeptis" (God has favored our undertakings); and
"Preserve me, O God: for in thee do I put my trust."
—Psalm 16:1

In the Senate Chamber:
"Annuit Coeptis" (God has favored our undertakings);and *"In God We Trust"*

iv) **Thomas Jefferson Building inscriptions:**

Above the figure of *Religion:*

"WHAT DOES THE LORD REQUIRE OF THEE, BUT TO DO JUSTLY,AND TO LOVE MERCY, AND TO WALK HUMBLY WITH THY GOD."
Quoted from the Holy Bible, Micah 6:8

Above the figure of *Science:*

"THE HEAVENS DECLARE THE GLORY OF GOD;
AND THE FIRMAMENT SHEWETH HIS HANDIWORK".

Quoted from the Holy Bible, Psalms 19:1" [11]

11. War Memorials:

From the records of the American Battle Monuments Commission, the following are a few examples of the U.S. war memorials that contain the name of God or Biblical references:

i. **The Arlington National Cemetery**, Arlington, Virginia: On its Chaplains Monument, dedicated to military clergy who fell in the First World War, are inscribed the words "TO THE GLORY OF GOD AND THE MEMORY OF THE CHAPLAINS WHO DIED IN SERVICES OF THEIR COUNTRY" and "MAY GOD GRANT PEACE TO THEM AND TO THE NATION THEY SERVED SO WELL".

ii. **The National Memorial Arch**, Valley Forge, Pennsylvania: This monument includes the words "...THIS VALLEY IN THE SHADOW OF THAT DEATH... ", which is based on Psalm 23, verse 4.

iii. **The Liberty War Memorial**, Kansas City, Missouri: It was built to honor the fallen American soldiers of the First World War. Its inscription includes the statement "THEIR BODIES RETURN TO DUST", which is a reference to Genesis, 3:19.

iv. **The American Military Cemetery**, Manila, Philippines: It contains the largest number of war graves of U.S. military dead (17,202) who were casualties of the Second World War.

There are also names of 36,286 military personnel missing in action inscribed in special walls. The memorial bears the following inscription: "GRANT UNTO THEM O LORD ETERNAL REST WHO SLEEP IN UNKNOWN GRAVES".

v. **The Cambridge American Cemetery**, Cambridge, England: This is the resting place of fallen Americans from the First and Second World Wars. It contains a memorial with the inscription "HERE LIES IN HONORED GLORY A COMRADE IN ARMS KNOWN BUT TO GOD".

vi. **The Normandy American Cemetery**, Normandy, France: This is a final resting place for American military personnel fallen in both the First and Second World Wars. A memorial there bears the inscriptions "HERE RESTS IN HONORED GLORY A COMRADE IN ARMS KNOWN TO GOD" and "MINE EYES HAVE SEEN THE GLORY OF THE COMING OF THE LORD".

vii. **The Brookwood American Cemetery and Memorial**, Brookwood, England: This location honors Americans who lost their lives in the First World War. An inscription there reads "HERE RESTS IN HONORED GLORY AN AMERICAN SOLDIER KNOWN BUT TO GOD…".

viii. **The Florence American Cemetery and Memorial** near Florence, Italy: Dedicated to U.S. military personnel lost in the Second World War, it has a panel which bears the inscription "…O GOD WHO ART THE AUTHOR OF PEACE . . ."

God made every one of us and I am sure it hurts Him when we kill each other. However, He does understand defense against evil.

12. Leaders' Prayers:

In times of regional, national, or international distress, it is noticeable to hear in person and through media, presidents provide prayers for the hurting.

Other political leaders, such as governors and mayors, have prayed for God's help in times of suffering, pain, and grief. We see and hear this repeatedly in the media, whether it is part of disaster relief, or in the aftermath of shootings, or other violent acts.

Many of these leaders have prayed for God's guidance in their work of governing. Unfortunately, not all leaders do this.

13. Leaders' Bible Studies:

On Capitol Hill in Washington, DC there are bible studies, such as the Congressional Prayer Caucus, for those elected members who wish to attend.

14. U.S. Public Holidays intended for honoring God:

The major public holidays in the U.S. intended for honoring God are Good Friday, Easter Sunday, Thanksgiving, and Christmas.

i. **Good Friday:** This major public holiday is celebrated the Friday before Easter Sunday. The week leading up to Good Friday is sometimes celebrated as Holy Week. It commemorates the crucifixion of Jesus Christ as His sacrificial payment of the penalty for our spiritual sins. Spiritual redemption is a gift we only have to accept in order to receive.

ii. **Easter Sunday:** is a day to celebrate the resurrection of Jesus Christ after his death. It is the "Third Day" on which Jesus rose from the dead, as described in the New Testament. Easter Monday is also a holiday, but it does not have the significance of the immediately preceding days.

iii. **Thanksgiving Day:** President George Washington issued a Proclamation in 1789 to declare a time for official thanks to God. It read as follows:

"By the President of the United States of America – A Proclamation

Whereas it is the duty of all Nations to acknowledge the providence of Almighty God, to obey his will, to be grateful for his benefits, and humbly to implore his protection and favor—and Whereas both Houses of Congress have by their Joint Committee requested me to recommend to the People of the United States a day of public thanksgiving and prayer to be observed by acknowledging with grateful hearts the many signal favors of Almighty God, especially by affording them an opportunity peaceably to establish a form of government for their safety and happiness. . .

Given under my hand at the City of New York the third day of October in the year of our Lord 1789, Geo. Washington." [12]

Days of thanksgiving and praise to God for His provisions of foods and other blessings have been celebrated since the days of the pilgrims and other early colonists.

In the U.S., Thanksgiving Day is now celebrated on the fourth Thursday of November each year. One of this major holiday's traditions is to have a special meal with family and/or friends, and give thanks to God for His harvest foods and other blessings.

iv. **Christmas**: This is an annual commemoration of the birth of Jesus Christ, the central figure in the Godhead or Holy Trinity of Father, Son, and Holy Spirit. Christmas is a widely observed holiday celebrated on December 25 by multi-millions of people around the world.

The above examples are just a small sampling of God's involvement with the government of the United States. There is much more regarding His reference, respect, and influence federally, as well as all His involvement in state and county affairs. It is reasonable, logical, and important that citizens, including students, be allowed to be taught who the God of their nation *is,* as well as what he does for each one of us every day.

15. Current Believers in God in the U.S.A.:

The following excerpts indicate the results from a May 5-8, 2011 Gallup Poll on the subject of belief in God by U.S. citizens.[13] According to the poll, more than 9 in 10 Americans still say "Yes" when asked the basic question, "Do you believe in God?"
The following graphic notes are excerpts from the poll:

"Gallup Poll, May 5-8, 2011:

PRINCETON, NJ -- *More than 9 in 10 Americans still say 'yes' when asked the basic question 'Do you believe in God?';* this is down only slightly from the 1940s, when Gallup first asked this question.

Do you believe in God?

	Yes	No	No opinion
	%	%	%
2011 May 5-8 ^	92	7	1
1967 Aug 24-29	98	1	*
1965 Nov	98	2	1
1954 Nov 11-16	98	1	1
1953 Mar 28-Apr 2	98	1	*
1947 Nov 7-12 †	94	3	2
1944 Nov 17-22 ‡	96	1	2

* Less than 0.5%

^ Asked of a half sample

† WORDING: Do you, personally, believe in God?

‡ WORDING: Do you, personally, believe in a God?

GALLUP

Earlier in this book we provided many solid, scientific reasons for this belief.

Chapter 10
God in the Government of the U.K.

Alice-photo/Shutterstock.com

PALACE OF WESTMINISTER - U.K. PARLIAMENT

So, Who IS This 'God' of Our Nation? This chapter and the science chapters show many reasons why students in the UK have the inalienable right to be taught about Him. He is a significant part of their government and can provide wisdom, hope, encouragement, grace, understanding, and more accurate knowledge of life.

Shown in this chapter are a few of the areas where God in the U.K. Government has been acknowledged in:

- the monarchy;
- the courts and justice systems;
- the constitution;
- the national anthem;
- the Prime Ministers' speeches;
- the currency;
- national buildings and places;
- government publications;
- war memorials;
- national holidays;
- believers.

The following examples of God's recognition show His importance to the nation:
(The following quotes from U.K. National Archives contain public sector information licensed under the Open Government Licence v2.0. This does not infer an endorsement of its use herein).

Note: Author emphasis by underlining.

1. God's Involvement with some of the United Kingdom's Monarchs

i. The Coronation Oath of King William and Queen Mary

"The *Coronation Oath* Act of 1688, when employed at Coronations, uses the <u>King James Bible.</u>
"An Act for Establishing the Coronation Oath.
"Oath heretofore framed in doubtful Words.

"Whereas by the Law and Ancient Usage of this Realme the Kings and Queens thereof have taken a Solemne Oath upon the Evangelists at Their respective Coronations to maintaine the Statutes Laws and Customs of the said Realme and all the People and Inhabitants thereof in their Spirituall and Civill Rights and Properties But forasmuch as the Oath itselfe on such Occasion Administred hath heretofore beene framed in doubtfull Words and Expressions with relation to ancient Laws and Constitutions at this time unknowne To the end therefore that One Uniforme Oath may be in all Times to come taken by the Kings and Queens of this Realme and to Them respectively Adminstred at the times of Their and every of Their Coronation.

"Annotations: Modifications etc. (not altering text)

"CI Short title given by Statute Law Revision Act 1948 (c. 62), Sch. 2

"II: Oath hereafter mentioned to be adminstered, by the Archbishop of Canterbury, & c.E+W

"May it please Your Majesties That the Oath herein Mentioned and hereafter Expressed shall and may be Adminstred to their most Excellent Majestyes King William and Queene Mary whome God long preserve at the time of Their Coronation in the presence of all Persons that shall be then and there present at the Solemnizeing thereof by the Archbishop of Canterbury or the Archbishop of Yorke or either of them or any other Bishop of this Realme whome the King's Majesty shall thereunto appoint and who shall be hereby thereunto respectively Authorized which Oath followeth and shall be Administred in this Manner That is to say,

"III: Form of Oath and Adminstration thereof.E+W

"The Arch-Bishop or Bishop shall say,

"Will You solemnely Promise and Sweare to Governe the People of this Kingdome of England and the Dominions thereto belonging according to the Statutes in Parlyament Agreed on and the Laws and Customs of the same?

"The King and Queene shall say,

"I solemnly Promise soe to doe.

"Arch Bishop or Bishop,

"Will You to Your power cause Law and Justice in Mercy to be Executed in all Your Judgements.

"King and Queene,

"I will.

"Arch Bishop or Bishop.

"Will You to the utmost of Your power Maintaine the Laws of God the true Profession of the Gospell and the Protestant Reformed Religion Established by Law? And will You Preserve unto the Bishops and Clergy of this Realme and to the Churches committed to their Charge all such Rights and Priviledges as by Law doe or shall appertaine unto them or any of them.

"King and Queene.

"All this I Promise to doe.

"After this the King and Queene laying His and Her Hand upon the Holy Gospells, shall say,

"King and Queene

"The things which I have here before promised I will performe and Keepe Soe help me God.

"Then the King and Queene shall kisse the Booke.

"IV: Oath to be adminstered to all future Kings and Queens. E+W

"And the said Oath shall be in like manner Adminstred to every King or Queene who shall Succeede to the Imperiall Crowne of this Realme at their respective Coronations by one of the Archbishops or Bishops of this Realme of England for the time being to be thereunto appointed by such King or Queene respectively and in the Presence of all Persons that shall be Attending Assisting or otherwise present at such their respective Coronations Any Law Statute or Usage to the contrary notwithstanding." [1]

ii. **The Coronation Oath from the Coronation of Elizabeth II, June 2, 1953**

The Archbishop of Canterbury conducted this oath, administering it in the form of questions:

" *Archbishop*: Will you solemnly promise and swear to govern the Peoples of the United Kingdom of Great Britain and Northern Ireland, Canada, Australia, New Zealand, The Union of South Africa, Pakistan, and Ceylon, and of your Possessions and the other Territories to any of them belonging or pertaining, according to their respective laws and customs?

"*Queen*: I solemnly promise so to do.

"*Archbishop*: Will you to your power cause Law and Justice, in Mercy, to be executed in all your judgements?

"*Queen*: I will.

"*Archbishop*: Will you to the utmost of your power maintain the Laws of God and the true profession of the Gospel? Will you to the utmost of your power maintain in the United Kingdom the Protestant Reformed Religion established by law? Will you maintain and preserve inviolably the settlement of the Church of England, and the doctrine, worship, discipline, and government thereof, as by law established in England? And will you preserve unto the Bishops and Clergy of England, and to the Churches there committed to their charge, all such rights and privileges, as by law do or shall appertain to them or any of them.

"*Queen*: All this I promise to do.

"Then the Queen, arising out of her chair, supported as before, the Sword of State being carried before her, goes to the Altar to makes her solemn oath in the sight of all the people to observe the premisses laying her right hand upon the Holy Gospel in the great Bible (which was before carried in the procession and is now brought from the Altar by the Archbishop, and tendered to her as she kneels upon the steps), and saying these words:

"*Queen*: The things which I have here before promised, I will perform and keep. So help me God.

"Then the Queen kisses the Bible and signs the Oath." [2]

2. Judicial Oaths

"When judges are sworn in they take two oaths/affirmations. The first is the oath of allegiance and the second the judicial oath; these are collectively referred to as the judicial oath.

i. **Oath of Allegiance**

"I, __________, do swear by Almighty God that I will be faithful and bear true allegiance to Her Majesty Queen Elizabeth the Second, her heirs and successors, according to law." [3]

ii. **Judicial Oath**

"I, __________, do swear by Almighty God that I will well and truly serve our Sovereign Lady Queen Elizabeth the Second in the office of ________ , and I will do right to all manner of people after the laws and usages of this realm, without fear or favour, affection or ill will." [4]

3. Parliamentary Oaths

"In the House of Commons, after election, an MP must swear an Oath of Allegiance before taking his or her seat. Members who object to swearing an oath may make a Solemn Affirmation instead.

"In the House of Lords the Oath of Allegiance must be taken, or Solemn Affirmation made, by every Lord on introduction and at the beginning of every new Parliament. This must be done before he or she can sit and vote in the House of Lords.

"While holding a copy of the New Testament (or, in the case of a Jew or Muslim, the Old Testament or the Koran) a Member swears: "I...swear by Almighty God that I will be faithful and bear true allegiance to Her Majesty Queen Elizabeth, her heirs and successors, according to law. So help me God."

"The text of the affirmation is: – "I...do solemnly, sincerely and truly declare and affirm that I will be faithful and bear true allegiance to Her Majesty Queen Elizabeth, her heirs and successors according to law". [5]

"In the United Kingdom, a significant part of the formal State Opening of Parliament is a Speech from the Throne. This is currently made by Her Majesty, Queen Elizabeth II. The Speech outlines the main bills to be introduced in the session of parliament and concludes with the statement, "My Lords and Members of the House of Commons, I pray that the blessing of Almighty God may rest upon your counsels." [6]

4. Parliamentary Sovereignty and the UK Constitution

"Parliamentary sovereignty is a principle of the U.K. constitution. It makes Parliament the supreme legal authority in the U.K., which can create or end any law. Generally, the courts cannot overrule its legislation and no Parliament can pass laws that future Parliaments cannot change. Parliamentary sovereignty is the most important part of the U.K. constitution.

"People often refer to the U.K. having an 'unwritten constitution' but that's not strictly true. It may not exist in a single text, like in the U.S.A. or Germany, but large parts of it are written down, much of it in the laws passed in Parliament – known as statute law.

"Therefore, the U.K. constitution is often described as 'partly written and wholly uncodified'. (Uncodified means that the U.K. does not have a single, written constitution.) …." [7]

Therefore, the various statutes that recognize God, are part of the constitution.

5. National Anthem

GOD SAVE THE QUEEN

God save our gracious Queen,
Long live our noble Queen,
God save the Queen!
Send her victorious ,
Happy and glorious,
Long to reign over us;
God save the Queen!
Thy choicest gifts in store
On her be pleased to pour;
Long may she reign;
May she defend our laws,
And ever give us cause
To sing with heart and voice,
God save the Queen! [8]

6. Prime Ministers' Speeches wherein God was Recognized

Many British Prime Ministers including Churchill, Thatcher, Brown, Cameron, and others have made statements regarding Britain's reliance on God's help with the survival, success, and well-being of the United Kingdom.

7. Currency

Today, most U.K. coins recognize God through the inscription "DEI GRATIA REGINA" or an abbreviation, "DEI GRA REG", or "D. G. REG". These all mean, "By the Grace of God, Queen" (or King). Another term that appears on the currency is "FIDEI DEFENSOR," "FID DEF," or "F.D." which all mean "Defender of the Faith."

This is another recognition of God by the government of the United Kingdom.

8. National Buildings and Places

Within the United Kingdom many public buildings bear inscriptions or artistic works that acknowledge God. (We address war memorials separately, later in this chapter).

The following are examples of a few inscriptions in Westminster Abbey:

i. On a processional cross: "Nation shall not lift up sword against nation, neither shall they learn war anymore." From the Bible, Isaiah 2:4. [9]

ii. On an Abbey bell: "Christie Audi Nos" which translated means "Christ Hear Us" [10]

iii. In the south aisle of the nave, on a monument to Carola Morland, second wife of Sir Samuel Morland, a Gentleman of the Privy Chamber, "Blessed be thou of the Lord, my honoured wife! Thy memory shall be a blessing, O virtuous woman" [11]

iv. Below each dial of "Big Ben" on the Elizabeth Tower, Palace of Westminster, the following inscription carved in stone: "Domine Salvam fac Reginam nostrum Victoriam primam" which means "O Lord, save our Queen Victoria the First." [12]

9. God in Government Publications

Take as an example the "Companion to the Standing Orders and Guide to the Proceedings of the House of Lords"

The "Appendix K" in this official document contains choices of Biblical verses that can be read to the House of Lords, as well as the option of six possible prayers to be read for opening each sitting of the House. [13]

10. War Memorials

Bikeworldtravel/Shutterstock.com

WAR MEMORIAL AT SLOANE SQUARE, CHELSEA, LONDON

The **U.K. National Inventory of War Memorials (UKNIWM)** was founded in 1989 to build a comprehensive record of every war memorial (roughly 100,000) in the United Kingdom, (including England, Scotland, Wales, and Northern Ireland), plus the Isle of Man and the Channel Islands. The name has since been changed to the Imperial War Museums War Memorials Archive.

The following are just a few of the war memorials that bear inscriptions acknowledging God, besides honouring the heroes who gave their lives defending the people of their nation:

i. **The West Hartlepool War Memorial**, Hartlepool, Cleveland, England. Also known as the Victory Square War

Memorial or the Victory Square Cenotaph. It commemorates those who died in the First and Second World Wars and includes the inscription: "...Thine O Lord Is The Victory".

ii. **First World War Memorials**: These include Ayton War Memorial, in Ayton, Berwickshire; the Earlston War Memorial, in Earlston, Berwickshire; the Gordon War Memorial, in Gordon, Berwickshire; and the Whitsome War Memorial, in Whitsome, Berwickshire. All are inscribed *"To the glory of God ..."*. The Whitsome War Memorial includes a second plaque, honoring those who fell in the Second World War, which is also inscribed *"To the glory of God ..."*.

iii. **The Ford and Etal War Memorial,** *Ford, Northumberland*: It consists of two marble tablets: the first one is dedicated to the memory of those who fell in the First World War and reads *"Grant them, O Lord, eternal rest"*. The second tablet, immediately below it, is dedicated to those who died in the Second World War.

iv. **The Anglo-French War Memorial,** *Thiepval, Picardie, France*: It commemorates those who lost their lives in the Battle of the Somme or in the First World War. The British headstones are inscribed with the phrase: *"A Soldier of the Great War/Known to God"*.

v. **The Scottish National War Memorial**, *Edinburgh, Scotland*: includes the inscription *"Thanksgiving and Praise to God"* and *"To the Glory of God."*

vi. **The National Monument to the Women of World War II,** *Whitehall, London, England.* It contains a frieze with an inscription that concludes with the words *"Glory be to God on high and on earth peace"*.[14]

11. Prayers in Parliament

Both the House of Commons and the House of Lords open their sessions with prayers. These are read by the chaplain in the Commons and by the senior bishop in the House of Lords.

"The form of the main prayer (in Commons) is as follows: "Lord, the God of righteousness and truth, grant to our Queen and her government, to Members of Parliament and all in positions of responsibility, the guidance of your Spirit. May they never lead the nation wrongly through love of power, desire to please, or unworthy ideals but laying aside all private interests and prejudices keep in mind their responsibility to seek to improve the condition of all mankind; so may your kingdom come and your name be hallowed. Amen."' [15]

The senior bishop in the House of Lords can consult the "Companion to the Standing Orders" for a choice of prayers to read. (See point 10 "God in Government Publications" earlier in this chapter).

12. U.K. National Holidays in God's Honour

i. **Good Friday** is the Friday before Easter Sunday. The week leading up to Good Friday is sometimes celebrated as Holy Week. Good Friday is a major religious holiday. It commemorates the crucifixion of Jesus Christ as His sacrificial payment of the penalty for our spiritual sins. Spiritual redemption is a gift we only have to accept in order to receive.

ii. **Easter Sunday** (holiday taken on Easter Monday) is a day to celebrate the resurrection of Jesus Christ after his death. It is the "third day" of His sacrificial time, and is the day He arose from His dead state, as described in the New Testament and historically established.

iii. **Christmas** is an annual commemoration of the birth of Jesus Christ, the central figure in the Godhead or Holy Trinity of Father, Son, and Holy Spirit.

Christmas is a widely-observed holiday, generally celebrated on December 25 by multi-millions of people around the world. This is just a small sampling of God's involvement with the national government of the United Kingdom. There is much more regarding His recognition, respect, and influence nationally, as well as all His involvement at the county level. It is reasonable, logical, and important that all U.K. citizens, including students, be taught who this God of their nation *is* and what he does for each of them every second of every day.

Current Believers in God in the U.K.

The information here was gathered from "Census 2011, Key Statistics, KS209EW" regarding religions of the people of England and Wales.

The total number of citizens covered was 56,075,912, of which 37,940,651 claimed to have a religious belief.

KS209EW – Religion (2011):

England and Wales

	Value	Percent
All categories: Religion	56,075,912	100.0
Has religion	37,940,651	67.7
Christian	33,243,175	59.3
Buddhist	247,743	0.4
Hindu	816,633	1.5
Jewish	263,346	0.5
Muslim	2,706,066	4.8
Sikh	423,158	0.8
Other religion	240,530	0.4
No religion	14,097,229	25.1
Religion not stated	4,038,032	7.2 ”

Source: Office for National Statistics. [16]

In the 2011 Census in **Scotland**, of 5,295,000 covered, 2,850,000 (about 54%) indicated that they were Christian, while 2,309,000 indicated they either had no religion or did not answer the question regarding their religion. [17]

In the 2011 Census in **Northern Ireland**, of the 1,720,645 enumerated, 1,424,896 (about 82.8%) claimed to be Christian, while 281,907 (about 16.4%) claimed no religion or did not answer the religion question. [18]

Chapter 11
God in the Government of Australia

by Dr. Graham McLennan

Dan Breckwoldt/Shutterstock.com

PARLIAMENT HOUSE, CANBERRA

So, Who IS This 'God' of Our Nation? This chapter and the science chapters show many reasons why students in Australia have the inalienable right to be taught about Him. He is a significant part of their government and can provide wisdom, hope, encouragement, grace, understanding, and more accurate knowledge of life.

(Note: Editor emphasis added by underlining).

European Discovery and Settlement of Australia

Two adventurous, European navigators, Christopher Columbus in 1492 and Ferdinand Magellan in 1519, sailed west across the Atlantic Ocean in search of a shorter route to India for trading. Both believed that God guided them and protected them on these dangerous journeys of discovery. Columbus discovered the Americas and Magellan was the first navigator whose ship and surviving crew were the first to sail around the world. Unfortunately, Magellan was killed in the Philippines. Neither adventurer found Australia but they showed a new way for others.

The western side of Australia was discovered by sailors such as William Dampier. Although he was a confirmed rogue, Dampier, stated in the preface to his book, *A Voyage to New Holland, an English Voyage of Discovery to the South Seas in 1699*, "But this Satisfaction I am sure of having, that the Things themselves in the Discovery of which I have been employed, are most worthy of our diligentest (sic) Search and Inquiry; being the various and wonderful Works of God in different Parts of the World." [1]

However, it was not until 1770 that Captain James Cook discovered the east coast of Australia and claimed it for Great Britain.

Matthew Flinders, who had the honour of naming Australia, was the first to circumnavigate the continent in 1802/1803 with this goal, "... to make so accurate an investigation of the shores of Terra Australis that ...with the blessing of God, nothing of importance would be left for future discoverers upon any part of these extensive coasts." [2]

European Settlement

"Governor Arthur Philipp arrived from England with the first fleet in 1788 to settle Australia with soldiers and criminals who could no longer be transported to North America because of American Independence.

"His instructions were to "*enforce a due observance of religion and good order among the inhabitants, and take such steps for the due celebration of public worship as circumstances would permit. In the first draft of these instructions he was to grant full liberty of conscience, and the free exercise of all modes of religious worship not prohibited by law, provided his charges were content with a quiet and peaceable enjoyment of the same, not giving offence or scandal to government; he was to cause the laws against blasphemy, profaneness, adultery, fornication, polygamy, incest, profanation of the Lord's Day, swearing and drunkenness to be rigorously executed. He was not to admit to the office of justice of the peace any person whose ill-fame or conversation might occasion scandal; he was to take care that the Book of Common Prayer as by law established be read each Sunday and Holy Day, and that the Blessed Sacrament_be administered according to the rites of the Church of England. Because of the great disproportion of female to male convicts, he was to take on board at any of the islands any women who might be disposed to come, taking care not to make use of any compulsive measures or fallacious pretences. He was to emancipate from their servitude any of the convicts who should, from their good conduct and a disposition to industry, be deserving of favour, and to grant them land, victual them for twelve months and equip them with tools, grain, and such cattle, sheep and hogs as might be proper, and could be spared. As the military officers and others might be disposed to cultivate the land, he was to afford them every encouragement."* [3]

Other nations could have settled Australia but mostly their beliefs prevented them from doing so: Hindus prevented sea voyages and contact with foreigners; a revolution in China in 1433 ended the voyages of navigator Cheng Ho; before the 1400s, Muslim sailors believed the southland was Dedjdal or the kingdom of Antichrist; and European expansion had begun in the East Indies and Pacific ending the expansion of Islam.

Early Leadership in Australia

Most of the colonies' early leadership came from the evangelical Christian community, mainly chaplains. Governors such as Hunter, Macquarie, and Brisbane, and a number of officials such as Judge Advocates Wylde and Ellis Bent, the editor of Australia's first newspaper, were strongly committed to Christian views, as were the school teachers.

Governor Macquarie was always trying to improve the moral and religious well-being of the colony, hoping that those in his care would become good Christians. He personally promoted the British and Foreign Bible Society and the Sunday School Movement. He also encouraged other Christian groups such as the Auxiliary Bible Society, and spoke at the Inaugural meeting. Macquarie particularly encouraged Christian education starting a number of schools under the supervision of the government chaplains so that by 1817 the most common discussion in the pages of the Sydney Gazette was on the merits of Bible reading.

James Stephen, the Permanent Under Secretary of the Colonial Office, believed that God was going to sovereignly use Australia as a Christian nation. He thought that it should be governed by Biblical principles, and encouraged Christian families to settle here. Hence, he was influential in the choice of Christian leaders in the colonizing of the country.

Australia's modern education system was pioneered by the chaplains and remained overtly Christian up to the 1880s. After that, it became more secular with the Anglican Church aligning with independent denominations in an unsuccessful attempt to stop the influence of Catholicism.

In 2011, the federal government increased funding for chaplaincy in public schools. That same year, attempts to remove the date designations "BC" (*Before Christ*) and "AD" (*Anno Domini*, Latin for "in the year of our Lord") from the national history curriculum were defeated.

South Australia's Godly Beginnings and the Aspirations of its Founders

For many years South Australia's capital, Adelaide, was known as the Holy City, but today it is called the City of Churches. In its formative years, Adelaide did not have enough churches for all of its parishioners.

During Adelaide's first eight years there were more preachers and places of worship than in the first decade of settled life in New England in the United States. From the time of South Australia's settlement in 1836 to 1915 more children attended Sunday school than attended regular school. In one of the first schools opened by Richard Angas, the sole textbook was the Bible. Angas distributed millions of gospel tracts in his lifetime.

Prayer and Meditation

Sir George Grey, a South Australian governor, shared with James Stephen in the Colonial Office, the view that "prayer and meditation on God's Holy Word ... were the inexhaustible, unfathomable source of all pure consolation and spiritual strength." [4]

Later, Grey was instrumental in the founding of New Zealand. New Zealand recently celebrated the bicentenary of the gospel arriving on its shores on Christmas Day, 1814, by the Reverend Samuel Marsden.

Godly Elements in Australia's Foundation

1. Law and Parliament

Australia's common law has been based on the Christian faith, exemplified by the statue of Jesus that occupied the central place above the Royal Courts of Justice in London, and many statements made by scholars. One Chief Justice declared: *"Christianity is parcel of the Common Law of England and therefore to be protected by it. So whatever strikes at the very root of Christianity tends manifestly to the dissolution of civil government."*

Australia's oldest parliament in New South Wales (NSW) governed most of Australia and many of the South Pacific islands including New Zealand. Today it still opens with this prayer: *"Almighty God, we ask for your blessing upon this Parliament. Direct and prosper our deliberations to the true welfare of Australia and the people of New South Wales. Amen."* Parliament of New South Wales Standing Orders 39. [5]

A similar prayer is said in our Federal Parliament by the President, who, upon taking the chair each day, reads the following prayer: *"Almighty God, we humbly beseech Thee to vouchsafe Thy special blessing upon this Parliament, and that Thou wouldst be pleased to direct and prosper the work of Thy servants to the advancement of Thy glory, and to the true welfare of the people of Australia.*

"Our Father, which art in Heaven, Hallowed be Thy name. Thy kingdom come. Thy will be done in earth, as it is in Heaven. Give

us this day our daily bread. And forgive us our trespasses, as we forgive them that trespass against us. And lead us not into temptation; but deliver us from evil: For thine is the kingdom, and the power, and the glory, for ever and ever. Amen." [6]

Alfred Deakin, the man mainly responsible for the passage of the Australian Constitution through the English House of Commons, repeatedly prayed over this significant document.

During this time, just prior to the turn of the century, Christians were coming together to discuss the federation movement (the alignment of separate states into one country). Many wanted to see God recognized as the ruler of the nation. Hence, it was carried unanimously in the Constitutional Convention that the preamble to Australia's Constitution should include the phrase, *"...humbly relying on* *the blessing of Almighty God."* Deakin was delighted with this outcome. (See next item, Our 1901 Commonwealth Constitution).

Deakin became Australia's second prime minister, after Edmund Barton, whose Presbyterian minister, Dr. Robert Steele, had inspired him to enter politics. The fourth prime minister, Sir George Reid, was also inspired to enter politics through Dr. Steele's influence. Deakin, who was born in Australia, was nurtured in his faith by his mother. It was Deakin who seconded the motion put forth by "Father of Federation" Sir Henry Parkes, for the proposed Federation of the Australian States.

Deakin kept a spiritual diary and from 1884 to 1913 wrote a *Boke of Praer and Praes* which contained nearly four hundred prayers. They mainly related to major decisions in his public life, revealing his utter dependence on God.

In the concluding words of his book *The Federal Story,* Deakin remarks that the Federation and the Australian Constitution were "providential" and were secured only "by a series of miracles".

In his notes in 1905 Deakin remarks, *"Sufficient to say that the religion of Jesus Christ is the life of the present, the light of the future and the hope of the world."* Many years later he stated: *"A life, the life of Christ, that is the one thing needful – the only revelation required is there... we have but to live it."* [7]

A Christian statesman, Deakin was the first Attorney General of the Commonwealth, and as such, founder of the High Court of Australia. He served three times as prime minister when a considerable amount of the Commonwealth's initial legislation started. As prime minister he founded the Arbitration Court, and the Australian Navy, as well as choosing Canberra as the nation's capital.

On June 3rd, 1898 a polling day was held in N.S.W., Victoria, and Tasmania to vote on creating a federation. By midnight Deakin knew that Victoria had approved the bill by an overwhelming majority, that Tasmania had done likewise, but that the majority in New South Wales had not reached the minimum number required for the adoption of the Bill. Hence, Deakin prayed, *"Father of Nations, receive our psalm of thanksgiving. Enable us to pursue the cause of unity in spite of the obstacles which at present appear to beset our path elsewhere. Guide us to appeal to that which is best and purest so as to make its development and mastery sure under our forms of government. Aid us to purify ourselves by our labours for the general weal and to invoke spiritual and moral principles so as to link us with our brethren on the highest plane to which we can at present attain. God preserve this people and grant its leaders unselfish fidelity and courage to face all trials for the sake of brotherhood. Thy blessing has rested upon us here yesterday and we pray that it may be the means of creating and fostering throughout all Australia a Christ-like citizenship."* [8]

2. The 1901 Commonwealth Constitution

The Preamble to the Australian Constitution states:

"Whereas the people of New South Wales, Victoria, South Australia, Queensland; and Tasmania humbly relying on the blessing of Almighty God, have agreed to unite in one indissoluble Federal Commonwealth under the Crown of the United Kingdom of Great Britain and Ireland, and under the Constitution hereby established..." These words are quoted from the Commonwealth of Australia Constitution Act, 1900. [9]

This preamble was in response to numerous signed petitions from people from every colony represented in the Federal Convention. This acknowledgement of the sovereignty of God was approved unanimously.

It is unlikely that Federation of the States would have been approved if the preamble had not included reference to Almighty God, as alluded to by Mr. Lyne (N.S.W.) in the debate. Some members of the constitutional convention had reservations, but with the inclusion of section 116, this was resolved and the "recognition insertion" was carried unanimously. Section 116 states:

"The Commonwealth shall not make any law for establishing any religion, or for imposing any religious observance, or for prohibiting the free exercise of any religion, and no religious test shall be required as a qualification for any office or public trust under the Commonwealth." These words are quoted from the Commonwealth of Australia Constitution Act, 1900. [10]

3. Monarchy

Our Constitutional Christian Monarchy expresses the Lordship of Christ when the Queen is presented with the Bible: *"...to keep your Majesty ever mindful of the law and the Gospel of God as the rule for the whole of life and government of Christian Princes, we present you with this Book, the most valuable thing this world affords. Here is wisdom; this is the royal law; these are the lively oracles of God."* [11]

When the royal orb is delivered to the Queen, the coronation service states: *"Receive this Orb set under the cross, and remember that the whole world is subject to the power and empire of Christ our Redeemer."* [12]

4. The Flag

Artgraphixel.com/Shutterstock.com

The Australian flag bears four Christian crosses. It incorporates the Southern Cross, which God has placed in the Southern Hemisphere, along with the crosses of St Andrew, St Patrick and St George.

5. Australian Days of Prayer

On June 11, 1738, John Wesley, a Christian theologian and Anglican cleric, blew the first trumpet call of the great evangelical revival. This was to have a deep and lasting effect on Britain and those in succeeding generations, prompting some to immigrate to Australian shores. Today, Wesley and his brother Charles are credited with founding the Methodist church movement.

Fifty years after the arrival of the first fleet, the NSW Governor, George Gipps – a Christian – proclaimed Sunday, November 2, 1838 a national day of fasting and humility because of severe drought. Within two days, heavy rains began to fall.

Almost six decades later on September 11, 1895 a day of prayer was again called in similar circumstances. Three weeks later a day of thanksgiving was proclaimed to thank God for the breaking of the drought.

Even the editorial in the April 14, 1897 edition of the Sydney Morning Herald stated: *"No Christian could in conscience vote for a Federation Bill that did not recognise God!"*

The first Sunday in the twentieth century was proclaimed Commonwealth Sunday and Christians were called to pray for the nation. During the 1940s the Second World War began to take its horrific toll and Australia was under threat, particularly after the bombing of Darwin. Several days of prayer were held; one of these was called by King George VI, to be held throughout the British Commonwealth.

Australia's biggest prayer meeting was held in 1988 during the country's bicentenary celebrations. Thirty-five thousand people surrounded the New Parliament House in Canberra, the nation's capital, representing triple the number who attended the official opening.

In 2004, the Governor-General of the Commonwealth of Australia, His Excellency Major General Michael Jeffery, fulfilled the desire of many Christians in Australia and launched a National Day of Thanksgiving. It was celebrated for the first time on May 29, 2004. In 2011, the inaugural National Day of Prayer and Fasting was held and in 2012 the inaugural National Christian Heritage Sunday took place on the first Sunday in February commemorating the first sermon on Australian soil on the 3rd day of February, 1788.

6. Opening of the Australian Parliament May 9, 1901- – A Christian Service

Royal Collection Trust / © Her Majesty Queen Elizabeth II 2014
This painting is on permanent loan to the parliament of Australia from the British Royal Collection. [13]

The opening of Parliament of the Commonwealth included the singing of Psalm 100, accompanied by an orchestra with His Excellency the Right Honourable the Earl of Hopetoun, a Member of His Majesty's Most Honourable Privy Council; Knight of the Most Ancient and Most Noble Order of the Thistle, Knight Grand Cross of the Most Distinguished Order of Saint Michael and Saint George, Knight Grand Cross of the Royal Victorian Order, Governor-General and Commander-in-Chief of the Commonwealth of Australia, reading the following prayers:

"O Lord, our heavenly Father, high and mighty, King of kings, Lord of lords, the only Ruler of princes, who dost from Thy throne behold all the dwellers upon earth, most heartily we beseech Thee with Thy favour to behold our most gracious Sovereign Lord King Edward, and so replenish him with the grace of Thy Holy Spirit that he may always incline to Thy will and walk in Thy way. Endue him plenteously with heavenly gifts, grant him in health and wealth long to live, strengthen him that he may vanquish and overcome all his enemies; and finally, after this life he may attain everlasting joy and felicity, through Jesus Christ our Lord. – Amen.

"Almighty God, the fountain of all goodness, we humbly beseech Thee to bless our gracious Queen Alexandra, George Duke of Cornwall and York, the Duchess of Cornwall and York, and all the Royal Family; endue them with Thy Holy Spirit; enrich them with Thy heavenly grace; prosper them with all happiness; and bring them to Thine everlasting Kingdom, through Jesus Christ our Lord. – Amen.

"Almighty God, we humbly beseech Thee to regard with Thy merciful favour the people of this land, now united in one Commonwealth. We pray for Thy servants the Governor-General,

the Governors of the States, and all who are or who shall be associated with them in the administration of their several offices.

"We pray Thee at this time to vouchsafe Thy special blessing upon the Federal Parliament now assembling for their first session, and that Thou wouldst be pleased to direct and prosper all their consultations to the advancement of Thy glory and to the true welfare of the people of Australia, through Jesus Christ our Lord, who has taught us when we pray to say:

"Our Father, which art in heaven, hallowed be Thy name. Thy kingdom come. Thy will be done in earth as it is in heaven. Give us this day our daily bread. And forgive us our trespasses, as we forgive them that trespass against us. And lead us not into temptation; but deliver us from evil; for Thine is the kingdom, and the power, and the glory, forever and ever. – Amen.

"The grace of our Lord Jesus Christ, and the love of God, and the fellowship of the Holy Ghost, be with us all, evermore. – Amen." [14]

The Australian Parliamentary Christian Fellowship conducts an annual Australian National Prayer Breakfast, under the oversight of the Parliamentary Chaplain in the Great Hall, in Parliament House, Canberra. It is attended by many dignitaries from Australia and overseas.

7. Currency

i. The Australian twenty-dollar note features Rev. John Flynn (1880-1951), who founded the Flying Doctor Service and the Australian Inland Mission. His Presbyterian ministers or Patrol Padres were known as the boundary riders of the bush, who rode camels to complete their mission work in central Australia. Flynn was responsible for using the pedal-wireless, a radio transmitter-receiver, to establish communication throughout inland Australia. This provided a new mantle of support and protection over the region, which was the size of Western Europe.

On the bottom-right-hand corner of the note is an image of one of the five camels that Flynn purchased in 1913 for his Patrol Padres.

ii. A picture of Pastor David Unaipon appears on the fifty-dollar note. An aboriginal, Unaipon was a writer, inventor, and pastor. A Church is visible on the bottom left corner of the bill. iii) An image of Caroline Chisholm (1808–1877) appeared on the five-dollar note for more than twenty years until 1992 when polymer notes were introduced. Married in the Church of England, she converted to her husband's religion of Catholicism.

She first arrived in New South Wales in 1838 then worked to establish better conditions, including suitable employment and accommodation, for young migrant women. Her work expanded to include making families' passage to Australia easier. What Australia needed most, in her view, were "good and virtuous women." In six years, she settled eleven thousand people as servants and farmers in NSW.

8. Swearing on the Bible in Court and Parliament

In courts, the Christian oath states: *"The evidence you shall give to the Court [and jury sworn] shall be the truth, the whole truth, and nothing but the truth, so help you God. Say 'I swear.' The witness then says, 'I Swear.'"*

Most people still swear on the Bible rather than make an affirmation, according to a criminal defense lawyer in Adelaide.

Most parliamentarians, prime ministers, and cabinet ministers are sworn in using the Bible. The Governor-General invites these officials to stand in their place and take the Bible in their right hand. The oath is read with the reply *"I do. So help me God!"* Prayers in local governments often appear in written form or are read by a clergy member.

9. Our Charitable Institutions

Thankfully, most of today's major Australian charitable and welfare agencies, such as the Salvation Army, Anglicare, Wesley Centre in Sydney and St. Vincent de Paul Society, continue as Christian institutions. Christian churches, more so than in England and the U.S., have been responsible for the development of social services.

10. Public Holiday – ANZAC Day on April 25th

Public holidays in Australia include Easter, Christmas, Australia Day, and the Queen's birthday. Throughout Australia and New Zealand, ANZAC (Australia New Zealand Army Corp) Day remembers our fallen soldiers in memorial services in virtually every town and suburb. They invariably are solemn Christian services such as the one led by a local pastor, held in Orange, N.S.W. in 2012, shown in the following photo. His public address included many Biblical references and concluded with a benediction.

Courtesy of Graham McLennan

11. ANZACS and Israel

The Australian and New Zealand armed forces commonly known as ANZACS have served in both the First and Second World Wars, as well as other theatres of war. One of the group's major achievements was the Australian Light Horse charge at Beersheba, on October 31, 1917 with eight hundred men on horseback. God was using one of the newest nations in the world to take Jerusalem from the Turks and Germans. This liberated Jerusalem from four hundred years of rule under the Turkish Ottoman Empire. Neither the military genius of Napoleon, nor the British Army with fifty thousand British Infantry who had fought bravely, had been successful in this challenge.

On the day of the Beersheba charge, the British government drafted the Balfour Declaration, which later became the foundation for the recognition of the State of Israel.

At every ANZAC service, the *Recessional Hymn*, written by the English poet and author, Rudyard Kipling in 1897, is sung:
God of our fathers, known of old
Lord of our far-flung battle line
Beneath Whose awful Hand we hold
Dominion over palm and pine
Lord God of Hosts, be with us yet,
Lest we forget, lest we forget. [15]
This hymn has become even more famous as the source of the oft-quoted phrase "*Lest we forget,*" used in ANZAC Day ceremonies.

Conclusion

We can see that Australia's discovery, settlement and growth can easily be explained in terms of God's intentions for our nation. He has used His men and women to lead in so many areas of development that even the most humanistic historian would have difficulty explaining away the mass of evidence at which this chapter only hints.

If the past is misinterpreted then so is the significance of the future. It is important that we don't continue to be deceived by the secularization process which denies the sovereignty of God in our history, past and present.

For further understanding of Australia's Christian heritage, please refer to the website of the Australian Christian History Research Institute (www.chr.org.au).

Current Believers in God in Australia:

The following excerpts were taken from the Australian Bureau of Statistics:

"2071.0 – Reflecting a Nation: Stories from the 2011 Census, 2012-2013.

(Latest ISSUE Released at 11:30 AM (CANBERRA TIME) 21/06/2012

"RELIGIOUS AFFILIATION

"Since the first Census, the majority of Australians have reported an affiliation with a Christian religion. However, there has been a long-term decrease in affiliation to Christianity from 96% in 1911 to 61% in 2011. Conversely, although Christian religions are still predominant in Australia, there have been increases in those reporting an affiliation to non-Christian religions, and those reporting 'No Religion'.

"In the past decade, the proportion of the population reporting an affiliation to a Christian religion decreased from 68% in 2001 to 61% in 2011. This trend was also seen for the two most commonly reported denominations. In 2001, 27% of the population reported an affiliation to Catholicism. This decreased to 25% of the population in 2011. There was a slightly larger decrease for Anglicans from 21% of the population in 2001 to 17% in 2011. Some of the smaller Christian denominations increased over this period

– there was an increase for those identifying with Pentecostal from 1.0% of the population in 2001 to 1.1% in 2011. However, the actual number of people reporting this religion increased by one-fifth." … [16]

From these numbers, we understand that at over 60% of Australia's population, those claiming to be Christians, believe in the God of our nation.

Editor's Note: This is just a sampling of God's involvement in the federal government of Australia. There are more examples regarding His recognition, respect, and influence federally, as well as in all the states, and other government areas. It seems reasonable, logical, and important that citizens, including students, should be allowed to be taught who the God of their nation *is* and what he does for each of us every second of every day.

Chapter 12
God in the Government of Canada

Songquan Deng/Shutterstock.com

THE PARLIAMENT BUILDINGS OF CANADA

So, Who IS This 'God' of Our Nation? This chapter and the science chapters show many reasons why students in Canada have the inalienable right to be taught about Him. He is a significant part of their government and can provide wisdom, hope, encouragement, grace, understanding, and more accurate knowledge of life.

(Note: Editor emphasis added by underlining).

Canada Becomes A Nation

The London Conference
December 1866 – March 1867

The Fathers of Confederation met in London, England from December 1866 to March 1867 to draft The British North America Act, Canada's first constitution.

The following notes are from Library and Archives Canada:

"Once New Brunswick and Nova Scotia had passed union resolutions in 1866 (the Province of Canada – later Ontario and Quebec – had already done so), it was time to meet to draft the text of the *British North America Act.* It was agreed that this meeting would take place in London. The Maritime delegates left for England on July 21, but for various reasons the Canadian delegation's arrival was delayed until late November. The conference was much smaller than those at Charlottetown or Québec had been, consisting of sixteen members in all (from New Brunswick, Nova Scotia, and the Province of Canada).

"After preliminary discussions, meetings officially began on December 4; they took place at the Westminster Palace Hotel in London. Business commenced with a thorough review of the Québec Resolutions to ensure that the wording of each was satisfactory. Despite Charles Tupper's promises to anti-union factions in Nova Scotia, he was unable to introduce amendments to the agreement at this time. Once the review was completed in late December, the "London Resolutions" were sent to the Colonial Office. Following the Christmas holiday, a committee of the delegates used the Resolutions to draft a proposed bill; copies were printed, and the delegates met with British officials in order to finalize the text.

“Choosing ‘Canada’ as the new country’s name was relatively easy, as was the choice of ‘Ontario’ and ‘Quebec’ for the two halves of the Province of Canada. However, difficulties arose in choosing a designation. The delegates wished it to be a kingdom; the British feared that such a title would anger the United States, and denied the request. An alternative, ‘Dominion,’ was suggested by Samuel Leonard Tilley, from a line in Psalm 72 of the Bible: ‘*He shall have dominion also from sea to sea, and from the river unto the ends of the earth.*’ (This was adopted).

“In addition to drafting the *British North America Act*, the Conference had to cope with the presence of an anti-union delegation from Nova Scotia, led by Joseph Howe, which was bent on overturning any union agreement. Charles Tupper was occupied in countering each submission Howe made to the Colonial Office and the two men conducted a debate through pamphlets and letters.

“The delegates had a completed text for the bill by the first week of February 1867. It was submitted to the Queen on February 11, and read in the House of Lords for the first time the following day. Proceedings were relatively uneventful: the bill passed through its first, second, and third readings in the House of Lords during the month of February. The three readings in the House of Commons were also swift, completed within two weeks with very little debate. The *British North America Act* received the Royal Assent on March 29, 1867.

“Once the Act was passed, the delegates returned home to prepare for union, which was scheduled to take place on July 1. Delegates from Nova Scotia and New Brunswick had to hold their final legislative sessions, in order to make last-minute changes to their constitutions. There also remained the task of selecting members for the new Cabinet and Senate.

"Social activities did not have the same prominence in London that they did at the other conferences, although some delegates did make excursions to other European countries, and visits to relatives and friends. For the most prominent of the delegates, there was also a royal audience. The major social event of the conference, however, was probably the marriage of John A. Macdonald (who became Canada's first Prime Minister) and Agnes Bernard on February 16, 1867." [1]

Reproduced with the permission of Rogers Communications Inc.

"THE FATHERS OF CONFEDERATION"
by Rex Woods 1969 [2]

Canada's Fathers of Confederation:

"1. Hewitt Bernard, secretary	14. Sir Alexander Campbell	26. Robert Barry Dickey
2. William Henry Steeves	15. Sir Adams George Archibald	27. Sir Charles Tupper

3. Edward Whelan	16. Sir Hector-Louis Langevin	28. John Hamilton Gray, N.B.
4. William Alexander Henry	17. Sir John Alexander Macdonald	29. William Henry Pope
5. Charles Fisher	18. Sir George-Etienne Cartier	30. William MacDougall
6. John Hamilton Gray, P.E.I.	19. Sir Étienne-Paschal Taché	31. Thomas D'Arcy McGee
7. Edward Palmer	20. George Brown	32. Andrew Archibald Macdonald
8. George Cole	21. Thomas Heath Haviland	33. Jonathan McCully
9. Sir Samuel Leonard Tilley	22. Sir Alexander Tilloch Galt	34. John Mercer Johnson
10. Sir Frederic B. T. Carter	23. Peter Mitchell	35. Robert Duncan Wilmot
11. Jean-Charles Chapais	24. James Cockburn	36. Sir William Pearce Howland
12. Sir Ambrose Shea	25. Sir Oliver Mowat	37. John William Ritchie
13. Edward Barron Chandler		

"This reproduction of Rex Wood's 1967 oil painting depicts the delegates to the 1864 Quebec Conference at which the basis of the British North America Act was formulated and discussed. The painting is based on an 1883 canvas by Robert Harris that was destroyed in the 1916 House of Commons fire. Confederation Life commissioned Rex Woods to recreate the Harris painting and it was presented to the nation as a centennial gift." [5]

God appears in a number of official documents and related content in the Government of Canada. Here are some key examples.

1. The Canadian Constitution (Excerpts)

"CONSTITUTION ACT, 1982

PART I

CANADIAN CHARTER OF RIGHTS AND FREEDOMS

Whereas Canada is founded upon principles that recognize the supremacy of God and the rule of law: (emphasis added)

Guarantee of Rights and Freedoms

1. The Canadian Charter of Rights and Freedoms guarantees the rights and freedoms set out in it subject only to such reasonable limits prescribed by law as can be demonstrably justified in a free and democratic society.

Fundamental Freedoms

2. Everyone has the following fundamental freedoms:

a. freedom of conscience and religion;

b. freedom of thought, belief, opinion and expression, including freedom of the press and other media of communication;

c. freedom of peaceful assembly; and

d. freedom of association.

Democratic Rights

3. Every citizen of Canada has the right to vote in an election of members of the House of Commons or of a legislative assembly and to be qualified for membership therein.

4. (1) No House of Commons and no legislative assembly shall continue for longer than five years from the date fixed for the return of the writs at a general election of its members.

(2) In time of real or apprehended war, invasion or insurrection, a House of Commons may be continued by Parliament and a legislative assembly may be continued by the legislature beyond five years if such continuation is not opposed by the votes of more than one-third of the members of the House of Commons or the legislative assembly, as the case may be." [3]

2. The Canadian Courts and Justice System

(Excerpts)

"Oaths of Allegiance Act

R.S.C., 1985, c. O-1

An Act respecting oaths of allegiance

SHORT TITLE

1. This Act may be cited as the *Oaths of Allegiance Act.*

R.S., c. O-1, s-1.

OATH OF ALLEGIANCE

2. (1) Every person who, either of his own accord or in compliance with any lawful requirement made of the person, or in obedience to the directions of any Act or law in force in Canada, except the *Constitution Act, 1867* and the *Citizenship Act*, desires to take an oath of allegiance shall have administered and take the oath in the following form, and no other:

I, ..., do swear that I will be faithful and bear true allegiance to Her Majesty Queen Elizabeth the Second, Queen of Canada, Her Heirs and Successors. So help me God. (Emphasis added.)

(2) Where there is a demise of the Crown, there shall be substituted in the oath of allegiance the name of the Sovereign for the time being.

R.S., c. O-1, s. 2; 1974-75-76, c. 108, s. 39.

SOLEMN AFFIRMATION

3. Every person allowed by law in civil cases to solemnly affirm instead of taking an oath shall be permitted to take a solemn affirmation of allegiance in the like terms, with such modifications

as the circumstances require, as the oath of allegiance, and that affirmation, taken before the proper officer, shall in all cases be accepted from the person in lieu of the oath and has the like effect as the oath.

R.S., c. O-1, s. 5. [4] .....

The National Anthems

i. ***The Royal*** (Current official lyrics)

"God save our gracious Queen,
Long live our noble Queen,
God save the Queen!
Send her victorious ,
Happy and glorious,
Long to reign over us;
God save the Queen!

Thy choicest gifts in store
On her be pleased to pour;
Long may she reign;
May she defend our laws,
And ever give us cause
To sing with heart and voice,
God save the Queen!" [5]

ii. **National Anthem** (excerpts as referenced)

The music for our national anthem was written in 1880 by Calixa Lavallée, known then as "Canada's national musician". The music was commissioned to go with the French words of a poem written by Judge Adolfe-Basile Routhie; the first performance was in Quebec City on June 24, 1880.

The unofficial English lyrics for Lavallée's music changed a number of times until the words of a poem written by Judge Robert Stanley Weir in 1908 were informally adopted as the English version of the anthem. These words were altered slightly until the first verse of his poem with the third-to-last line modified, was officially proclaimed as Canada's anthem through the National Anthem Act in 1980. The words are as follows:

O Canada! (English version)
O Canada! Our home and native land!
True patriot love in all our sons' command.
With glowing hearts we see thee rise,
The true north strong and free;
From far and wide, O Canada,
We stand on guard for thee.
God keep our land glorious and free!
O Canada, we stand on guard for thee,
O Canada, we stand on guard for thee.

Here is the balance of the original poem by Judge R. Stanley Weir in 1908:
O Canada! Where pines and maples grow,
Great prairies spread and lordly rivers flow,
How dear to us thy broad domain,
From east to western sea!
Thou land of hope for all who toil!
Thou true north strong and free!
O Canada! O Canada!

O Canada, we stand on guard for thee,
O Canada, we stand on guard for thee.

O Canada! Beneath thy shining skies
May stalwart sons and gentle maidens rise;
To keep thee steadfast through the years
From east to western sea,
Our own beloved native land,
Our true north strong and free.
O Canada! O Canada!
O Canada, we stand on guard for thee,
O Canada, we stand on guard for thee

Ruler supreme, who hearest humble prayer,
Hold our Dominion in thy loving care.
Help us to find, O God, in thee
A lasting rich reward.
As waiting for the better day,
We ever stand on guard.
O Canada! O Canada!
O Canada, we stand on guard for thee.
O Canada, we stand on guard for thee!

iii. ***O Canada!*** (French version):
O Canada! Terre de nos aïeux,
Ton front est ceint de fleurons glorieux!
Car ton bras sait porter l'épée,
Il sait porter la croix!
Ton histoire est une épopée
Des plus brillants exploits.

Et ta valeur, de foi trempée,
Protégera nos foyers et nos droits.
Protégera nos foyers et nos droits. [6]

4. Prime Ministers' and Their Party's Initial Throne Speeches (Excerpts)

The First Throne Speech was made to the Parliament of Canada on November 7, 1867 with the First Prime Minister, Sir John A. Macdonald in attendance. After the opening prayers, the Throne Speech was delivered by the First Governor General of Canada, His Excellency, The Right Honourable Charles Stanley:

".... I am happy to be able to congratulate you on the abundant harvest with which it has pleased Providence to bless you, and on the general prosperity of the Dominion. Your new nationality enters on its course backed by the moral support-the material aid-and the most ardent good wishes of the Mother Country. Within your own borders peace, security and prosperity prevail, and I fervently pray that your aspirations may be directed to such high and patriotic objects, and that you may be endowed with such a spirit of moderation and wisdom as will cause you to render the great work of Union which has been achieved, a blessing to yourselves and your posterity, and a fresh starting point in the moral, political and material advancement of the people of Canada." [7]

Recent Throne Speeches at the opening of new Parliament Sessions, delivered to the Senate and the House of Commons by the Governor-General, have concluded with these words:

"May Divine Providence guide you in your deliberations and make you equal to the trust bestowed upon you." [8]

5. Daily Proceedings in Parliament (Excerpts)

"Each of the three events in the Daily Proceedings—Prayers, Statements by Members, and Oral Questions—is covered separately in the Standing Orders.

i. **Prayers**

"Prior to the doors of the Chamber being opened to the public at the beginning of each sitting of the House, the Speaker takes the Chair and proceeds to read the prayer, after it has been determined that a quorum of 20 Members including the Chair Occupant is present, and before any business is considered. While the prayer is being read, the Speaker, the Members and the Table Officers all stand. The prayer is by custom read partly in French and partly in English. When the prayer is finished, the House pauses for a moment of silence for private thought and reflection. At the end of the moment of silence, the Speaker orders the doors opened. At this point, television coverage of the proceedings commences and the public may enter the galleries.

ii. **Historical Perspective**

. . . "Until 1994, no major change to the form of the prayer was made aside from references to royalty. At that time, the House concurred in a report recommending a new form of prayer more reflective of the different religions embraced by Canadians. This prayer was read for the first time when the House met to open its proceedings on February 21, 1994:

"Almighty God, we give thanks for the great blessings which have been bestowed on Canada and its citizens, including the gifts of freedom, opportunity and peace that we enjoy. We pray for our Sovereign, Queen Elizabeth, and the Governor General. Guide us in our deliberations as Members of Parliament, and strengthen us in our awareness of our duties and responsibilities as Members. Grant us wisdom, knowledge, and understanding to preserve the blessings of this country for the benefit of all and to make good laws and wise decisions. Amen.

"When the House convenes on the first day of a new Parliament or on any day when the House is to elect a Speaker, the prayer is read after a Speaker has been elected. Indeed, at that time, the election of a Speaker must be the first order of business and has precedence over all other matters. Only after a Speaker has been elected, is the House properly constituted to conduct its business. After the House reconvenes following the election of the Speaker, the prayer is read before the House proceeds to the Senate to inform the Governor General of its choice." [9]

6. Canadian War Memorials

i. The Tomb of the Unknown Soldier:

"The Tomb of the Unknown Soldier was created to honour the more than 116,000 Canadians who sacrificed their lives in the cause of peace and freedom. Furthermore, the Unknown Soldier represents all Canadians, whether they be navy, army, air force or merchant marine, who died or may die for their country in all conflicts – past, present, and future." [10]

TOMB OF THE UNKNOWN SOLDIER
Ottawa, Ontario

ii. Victory Square Cenotaph, Vancouver:

VICTORY SQUARE CENOTAPH
Vancouver, British Columbia [11]

This memorial bears the inscription "THEIR NAME LIVETH FOR EVERMORE."

The following excerpts are from a book, *Remembrance Day 1944: Service of the Armed Forces and Citizens, Vancouver, Canada,* written by Major J. S. Matthews and published by the City of Vancouver Archives:

"Those whose sacrifice this Cenotaph commemorates, were among the men who, at call of King and Country, left all that was dear, endured hardship, faced danger, and finally passed from the sight of men by the path of duty, giving their own lives that others might live in freedom. Let those who come after see to it that their names be not forgotten."

– Major the Reverend C.C. Owen

"The Cenotaph was unveiled by His Worship W.R. Owen, Mayor of Vancouver, in the presence of an assemblage of 25,000 persons, naval, military and civilian, and including The Old Contemptibles, 7th British Columbia, 29th Vancouver, 72nd Seaforth, 2nd Canadian Mounted Rifles, 47th New Westminster, and 102nd North British Columbia Battalions, C.E.F., and others, on Sunday, 27th April, 1924. It was dedicated by Hon. Major the Reverend Cecil C. Owen, M.B.E., V.D., D.D., Chaplain of the 29th (Vancouver) Battalion, C.E.F., *"To the Glory of God, and in thankful remembrance of those who served King and Country overseas in the cause of truth, righteousness and freedom."*

"The 24th Psalm was read by Hon. Lt.-Col. the Rev. G.O. Fallis, C.B.E., E.D., D.D., of the Methodist Church, and the music included "O Canada" (Buchan); "O God, Our Help in Ages Past", "Lochaber No More" (bagpipes); "For All the Saints"; "Last Post" and "God Save the King". The first wreath, being the tribute of the Corporation and Citizens of Vancouver, was reverently placed by Mrs. W.R. Owen, wife of His Worship, the Mayor.

J.S. Matthews,
City Archives,
City Hall, Vancouver, 1944." [12]

There are hundreds of various memorials across Canada in cities and towns, each honouring the brave souls who gave their lives in defense of Canadians, allies, and our precious democratic freedoms.

iii. **The poem, *In Flanders Fields*:**

"In Flanders Fields was first published in England's *Punch* magazine in December 1915. Within months, this poem came to symbolize the sacrifices of all who were fighting in the First World War. Today, the poem continues to be a part of Remembrance Day ceremonies in Canada and other countries throughout the world.

"The poem was written by a Canadian—John McCrae, a doctor and teacher, who served in both the South African War and the First World War.

"In Flanders Fields

In Flanders fields the poppies blow
Between the crosses, row on row,
That mark our place; and in the sky
The larks, still bravely singing, fly
Scarce heard amid the guns below.

We are the Dead. Short days ago
We lived, felt dawn, saw sunset glow,
Loved and were loved, and now we lie,
In Flanders fields.

Take up our quarrel with the foe:
To you from failing hands we throw

The torch; be yours to hold it high.
If ye break faith with us who die
We shall not sleep, though poppies grow
In Flanders fields.

By Lieutenant-Colonel John McCrae" [13]

7. National Holidays that Honour God

Of the five national holidays in Canada for everyone, plus five additional holidays for federal employees and most other citizens, four are established specifically to honour God and His Son, Jesus. The first five national holidays include New Year's Day, Good Friday (Easter), Canada Day, Labour Day, and Christmas Day. The other five are Easter Monday, Victoria Day, Thanksgiving Day, Remembrance Day, and Boxing Day.

The first of the God-honouring four national holidays is *Good Friday, which is established to honour the sacrificial death of Jesus Christ, the Son of God, who died to pay the penalty for the spiritual sins of all the people of the world who would recognize this as a gift of salvation and accept it.*

The second is *Easter Monday, which is established to honour the resurrection of Jesus from His dead state showing that God can give life to the dead physically as well as spiritually.*

The third of the four is *Thanksgiving Day, which is established to set aside a time to give special thanks to God for the harvest of the food He has made for us throughout the year plus all our other blessings.*

The fourth is *Christmas Day, which is established to honour the birth of Jesus Christ, Son of God. He was sent to teach us how to live a fulfilled and joyful life through loving our Creator and loving our neighbours as ourselves.*

God is also honoured frequently on Canada Day and Remembrance Day, in national programs and events, and in leaders' prayers.

Each time that we sing our national anthem, "O Canada", at sports events and on other occasions, we are requesting God to "keep our land glorious and free".

8. Canadian Currencies that Honour God

The Canadian dollar has various abbreviations, names, and nick-names including CAD, C, Dollar, Buck, or Loonie; one hundred cents to the dollar.

The Godly connection with Canadian currency is that all the coins currently bear the image of Queen Elizabeth II and the inscription "D. G. REGINA" which is short form of "DEI GRATIA REGINA", Latin for "BY GOD'S GRACE, QUEEN."

The Canadian dollar became the official currency for the Province of Canada (now Ontario and Quebec) in 1858.

After Confederation in 1867, the new Nation of Canada took control of the production of its currency. Coins are produced by the Royal Canadian Mint and the polymer bills are outsourced from the British American Bank Note Company.

Canada has an excellent credit rating and is internationally admired for its astute fiscal management.

9. Public Buildings and Places that Honour God

When approaching the Peace Tower at the centre of Canada's Parliament Buildings, you can see foundational principles of Canadian government inscribed thereon: *He hath dominion from sea to sea* (Psalm 72, v.8) and *Give the king thy judgements, O God, and thy righteousness unto the king's son* (Psalm 72, v.1).

Another verse inscribed on the Peace Tower is: *Where there is no vision, the people perish.* (Proverbs 29:18).

These indicate the type of wise government that Canada intended to provide. May our elected officials and government employees keep these concepts in mind.

Other Bible verses inscribed in Parliament's Memorial Chamber include Ephesians 6:13: *Wherefore take unto you the whole armour of God that ye may be able to withstand in the evil day, and having done all, to stand.*

Conclusion

The items and locations covered in this chapter are only a small sample of the honour, and respect shown by the federal government of Canada towards God and His involvement in governmental content. Much more information is available regarding His recognition and influence nationally, as well as all of His provincial and municipal involvement.

These inscriptions, and many others on public buildings, are there for the public to see and for guidance of our governments.

It is clear that God has been an influential help in the governance of Canada. His wisdom and guidance are always available and always beneficial to those of us who will consult Him by prayer and the study of His word, the Holy Bible.

Current Canadian Believers in God

The information here was gathered from "Statistics Canada's 2011 National Household Survey: Data Tables" regarding the Religion and Immigrant Status of Canadians. [14]
The total number of citizens covered was 32,852,320 of which 25,720,175 were Canadian citizens by birth and 6,775,765 were landed immigrants who had been granted the right to live permanently in Canada.

Of the total number covered by these statistics, 22,102,745, divided into 73 groups, consider themselves to be Christians; 2,898,970, divided into 26 groups, consider themselves to be part of other religions; and the other 7,850,605, divided into 5 groups, claim to have no religious affiliation.

One of Canada's "freedoms" is the freedom to choose what type of religion you want, including none at all.

From these numbers, we are given the impression that at least two-thirds of the population covered in this census, those claiming to be Christians, believe in the God of our nation.

Brief Summary of "Darwin's Replacement"

Some of us will remember the "Happy Days" when the God of our nation was a part of our schooling. There was an abundance of good clean fun, trust, and generally, a Godly moral standard that guided our actions and concern for each other.

Then one misguided woman named Madalyn Murray O'Hair with help from a few other atheists, managed to get the supreme court of the USA to ban prayers and Bible reading in government-funded schools. Soon most signs of God's presence, were removed. Other countries followed.

Then what began to happen almost immediately? Moral standards collapsed into wide-spread promiscuity with all its STDs, teen pregnancies, depressions, massive drug and alcohol addictions which are still growing in our nations today. All of this coupled with related escalating welfare costs, health-care costs, and policing costs at the expense of taxpayers, let alone all the heart-ache.

Our public-funded schools and secular universities ramped-up the teaching that Evolution-only is the cause of life (so what do we need a god for?).

The numbers of people suffering with depression, addictions, unhappy family relationships, and tragic moral standards are obvious all around us. It is as if the rumored conspiracy to bring democracies down from within, is working for the perpetrators.

Is this the best we can do to manage our nations?

Not by a long shot!

With the agreement and participation of concerned citizens, we can encourage our education leaders to allow our wise Creator to return to our classrooms, now for scientific reasons.

I. L. Cohen said it well:

"At that moment, when the RNA/DNA system became understood, the debate between Evolutionists and Creationists should have come to a screeching halt." [1]

(He is rightfully indicating that when the enormous and super-intelligent complexities of RNA and DNA Programming were discovered, the theory of evolution, which, by definition, has no intelligence to program with, should have been dropped right at that time).

"Any suppression which undermines and destroys that very foundation on which scientific methodology and research was erected, evolutionist or otherwise, cannot and must not be allowed to flourish. ...It is a confrontation between scientific objectivity and ingrained prejudice – between logic and emotion – between fact and fiction. ...In the final analysis, objective scientific analysis has to prevail – no matter what the final result is – no matter how many time-honored idols have to be discarded in the process....

It is not the duty of science to defend the theory of evolution, and stick by it to the bitter end – no matter what illogical and unsupported conclusions it offers... If in the process of impartial scientific logic, they find that creation by outside super-intelligence is the solution to our quandary, then let's cut the umbilical cord that tied us down to Darwin for such a long time. It is choking us and holding us back.

...Every single concept advanced by the theory of evolution (and amended thereafter) is imaginary as it is not supported by the scientifically established facts of microbiology, fossils, and mathematical probability concepts. Darwin was wrong. ... The theory of evolution may be the worst mistake made in science." [2]

With almost three decades of research, including substantial valuable input from 20 PhDs, 3 MDs, 9 DScs, 3 Mathematicians, 2 MScs, and 8 Independent Researchers, we believe that the resulting science of "Atomic Biology" is a logical replacement for Darwinisms, as the taught cause of life.

We have found to our amazement, that many scholars we have talked to, have never thought through what has to happen in building cells with atoms. When we asked the question, "How do YOU think the right numbers of the right atoms are selected from available resources and precisely placed to build all cell-parts?" the variety of answers were all so bizarre, we could tell that the highly educated individuals had never considered the capabilities, choices, decisions, super-intelligence, and dexterity required to do this work.

One of the reasons we know that it takes super-intelligence is that our scientists with all their accumulated knowledge and sophisticated equipment, cannot come anywhere close to building any of the molecular machines required for cells. Even the three scientists who won the 2016 Nobel Prize for Chemistry, Sauvage, Stoddart, and Feringa, in Europe, plus the James Tour Group at Rice University in Texas, just managed to build only some relatively simplistic molecular machines. This involved thirty years of research and development, sophisticated equipment, plus highly regulated and unnatural conditions.

Therefore, it is totally impossible for an unintelligent non-force like evolution to perform the brilliant construction work essential to build the phenomenally sophisticated molecular machines within our cells.

We have outlined several supported reasons why the construction, sustenance, maintenance, and repair of all the highly complex parts of our cells, requires super-intelligent physical work and care. This is work that evolution is incapable of performing as, by definition, it has no intelligence with which to do the work.

Consider the following:

The key components of DNA are 4 bases:

Adenine - chemical formula $C_5 H_5 N_5$

Guanine - " " $C_5 H_5 N_5 O_1$

Cytosine - " " $C_4 H_5 N_3 O_1$

Thymine - " " $C_5 H_6 N_2 O_2$

Notice how similar the formulae are for these bases. As each one is being constructed for us using atoms from our digestive system, it is so critical that the builder be absolutely precise in the selection decisions, choices, counting, and precision placement of the right numbers of the right atoms, as well as fastening the right bases in the right sequences in programming our DNA for the various required functions in our cells. What super-intelligence and care this takes. It certainly does not just happen as unguided chemical reactions.

Evolution is incapable of constructing and maintaining cells with all their molecular machines and other complex parts.

Evolution is incapable of providing the essential breath-of-life to the inanimate atoms used to build cell-parts.

Because Evolution lacks these essential capabilities, it is falsified as the theory of the cause of life.

Seven Basic Principles Of Life

Principle #*1*

Virtually all matter, living or not living, is constructed or built of atoms that do not have legs, brains, fins, or muscles. Since atoms have no internal means to move themselves into any precise position in a cell, a super-intelligent and capable external force is required to find, sort, select, count, grasp, and precisely place all of the right numbers of the right atoms necessary to build each cell of our food (fruit, vegetables, etc.); then, to perform a similar process with the right numbers of the right atoms from our digesting food and place them precisely to build each new complex cell for us; and then deliver each one of these cells to its precise position in our body, fasten it there, and hook it up properly to our blood vessels and nerve networks, etc.

Every step of this brilliant work must be performed with super-intelligent planning, care, dexterity, precision, and speed. In the English language, the name given this phenomenally intelligent, reliable, trustworthy, and caring super-intelligent force by citizens and Governments, is "God". (Evolution, by definition, has no intelligence to work with, therefore, it is **falsified** as the cause of life).

Principle #*2*

The super-intelligent physical work required to build living cells, goes far beyond the parameters of possibility for the unguided process theorized as Darwinism, Neo-Darwinism, or macro-evolution.

For example, over 4900 quadrillion right atoms per second must be found and sorted from our digesting food, then selected, precisely assembled into new red blood cells, and delivered to each average adult's blood stream just to supply his or her replacement red blood cells; that is for every adult every second of every day. [3] Of course, in the same second, a greater number of atoms (to make the leaves, roots, skin, etc. for our fruit and veggies) have to be found, sorted, and selected from the soil, water, and air of fields, gardens and orchards to be assembled into more food from which to take the right numbers of the right atoms to make each future second's red blood cells for each of us.

This is in addition to the phenomenal, reliable, caring physical work of constantly maintaining, repairing, and/or replacing the other approximately 80 trillion cells in each human adult body.

Evolution, by definition, has no intelligence to use, therefore, *it is incapable of doing any of the super-intelligent physical work* necessary to build each complex part of each cell. (Thus, Evolution is **falsified** as the cause of life).

Principle #*3*

"Dead dogs don't bark," which is to say that although all the right atoms, molecules, and cells are precisely built and placed into their correct position for its eyes, ears, teeth, brain, legs, heart, lungs, paws, liver, kidneys, stomach, fir, claws, nose, and so on, *without the super-intelligent breath-of-life, those atoms, molecules and cells are not going to move one millimeter.* This God-given breath-of-life is crucial to every living entity. When it is removed, the entity's life ends. (Evolution cannot provide this necessary attribute, therefore its theory as the cause of life, is **falsified**).

Principle #*4*

*Evolution is not a force. By definition, it has absolutely no intelligence, guidance, appendage, or plan to work with. However, all of these attributes plus super-intelligent vision, dexterity, precision, and speed are essential to construct, grow, maintain, repair, and care for each living entity. (As Evolution is not capable of performing these tasks, it cannot be the cause of life, and is therefore **falsified** as the cause of life).*

Principle #*5*

Mr. Darwin said his theory of evolution could "absolutely break down" *if the fossil recorders did not find examples of one kind of creature changing into another kind of creature. This would indicate that creatures do not evolve from one kind to another kind.* He knew there would need to be huge numbers of transition fossils. In his book, *On The Origin of Species,* he states, *"But just in proportion as this process of extermination has acted on an enormous scale, so must the number of intermediate varieties, which have formerly existed, be truly enormous. Why then is not every geological formation and every stratum full of such intermediate links? Geology assuredly does not reveal any such finely-graduated organic chain; and this, perhaps is the most obvious and serious objection which can be urged against the theory (of evolution).* [4]

Out of the millions of fossils found, there should be hundreds of thousands of transitions, but they do not exist. For example, there are no complete fossils of a man with a monkey's tail. Men and monkeys still exist, yet there is not even one complete transitional creature in existence or in the fossil record.

What we have found are minor changes within species such as different colors and sizes of humans, horses, cats, dogs, fish, birds, bugs, etc.

The occasional random mutations occurring between generations, that were at one time thought to be the cause of improvements in a species, actually involve a loss of DNA information. A more accurate word for these mutational changes is "devolution" or "degeneration," *never a change of one kind of entity to another kind.*

There have been many attempts to produce images of transitions from bone fragments and there are fossils that have some characteristics similar to other kinds of creatures. However, the understanding of the enormous amount of intelligent work involved in designing and building living entities with atoms, points far more logically to a common designer and builder rather than a common ancestor.

As stated by Dr. Gary Parker, a paleontologist, biologist, and former evolutionist, "Fossils are a great embarrassment to Evolutionary theory and strong support for the concept of Creation." [5] (The lack of transitions is a reassurance that Evolution is **falsified** as the cause of life).

Principle #*6*

All living entities are built of cells containing DNA. DNA is like a computer software program which has very sophisticated, highly complex coding to assist in the multiple and varied functions which cells have to perform in order to help keep a living entity alive.

Complex, intelligent, functional codes, like DNA, cannot be programmed without an intelligent programmer. (Therefore, Evolution is ***falsified*** *as the cause of life).*

Principle #*7*

There are many factors that must be super-intelligently tuned, highly regulated, and crucially consistent in order for our beautiful planet to function, and for living entities to exist.
Random, uncontrolled, inconsistent conditions would quickly lead to extinction of creatures.

Here are just a few conditions under which life would not exist:

1. If temperatures became too hot or too cold;
2. If there was insufficient water available;
3. If there was insufficient food;
4. If there was no one to build cells from atoms;
5. If there was no one to breathe life into these cells;
6. If there was no controlled sunlight;
7. If there was no controlled atmosphere;
8. If there was no controlled gravity;
9. If there was no controlled electricity;
10. If there was no super-intelligent force to keep all of these necessary factors (and more) in balance, life would not exist.

Fortunately for the creatures on our planet, we have a super-intelligent controller who reliably and consistently keeps everything necessary for life in balance. (As Evolution, by definition, has no intelligence to work with, it is ***falsified*** *as the cause of life).*

Here is a summary list of *essential, intelligent, physical works* required to build and maintain each living entity that *evolution cannot perform, as, by definition, it has no intelligence:*

1. Evolution cannot *see and think*;
2. It cannot *find* the necessary 'building-block' atoms;
3. It cannot *sort* the right atoms from the wrong ones for building each cell-part;
4. It cannot *count* the right numbers of each type of atom for building each cell-part;
5. It cannot *grasp and move* the right atoms for building each cell-part.
6. It cannot *precisely place* each atom to build each cell-part;
7. It cannot *fasten* each atom in its correct place in each cell-part;
8. It cannot *program* the RNA and DNA molecules as required for each cell;
9. It cannot *breathe life into* the inanimate atoms in each cell;
10. It cannot *re-sort* the atoms in eaten food to build our human cells;
11. It cannot *work quickly,* or at all;
12. It cannot *build* a living entity;
13. It cannot *sustain* the life of a living entity;
14. It cannot *maintain* any entity;
15. It cannot *repair* or heal any entity;

16. It cannot *build communication systems* within living entities;
17. It cannot *fine-tune our living conditions* on our planet;
18. It cannot *build blood-vessel networks* for creatures.

These are only a few of the super-intelligent works necessary for life on our planet, that evolution cannot perform.

For these and other reasons outlined herein, the theory of evolution as the cause of life, should be classified as **"falsified".**

So, what is the "Best Explanation" for the cause of life?

Some other theories partially involve evolution, e.g. Theistic Evolution, Evolutionary Creation, Progressive Creation, and variations thereof, theorize that God initiated atoms and natural laws, then just allowed life to originate and carry on with one kind of creature evolving into many other kinds of creatures descending as mutants from a common ancestor.

Reality R&D does not agree because our research revealed that every part of every cell in every living entity has to be intelligently constructed with atoms. There are super-intelligent decisions and choices necessary throughout the construction work.

Our common designer and builder, i.e. our Creator, replaces the idea of a common ancestor.

Intelligent Design is a good theory but it stops short of the whole story. If you ask any architect, engineer, manufacturer, building contractor, or kid with a Lego set, you will get the same answer: a design plan does not build the project. Intelligent physical work is essential.

After almost three decades of research, we are proposing that the Godly life-science of "Atomic Biology" be considered seriously as the "best explanation" for the cause of life.

Where it is accepted by various educators, it could be taught as the logical, evidence-based replacement for the currently taught theory of evolution.

We do not call this a "new" science because we believe our Creator has been using this since the beginning of life, along with the forces of gravity, electricity, magnetism, etc.

It would be very revealing to find how many teachers, professors, and scientists actually doubt evolution is the cause of life, if the threats and duress forcing the teaching and proclamation of evolution-only as the cause of life, were made illegal, as it should be.

According to Bob Enyart's research, his *"RSR's 2014 List(s) of Scholars Doubting Darwin"* includes seven lists of scientists "...who have gone out of their way to declare their doubt about Darwin...", plus websites showing that "...30,000 U.S. high school biology teachers who do not endorse Darwinism in class" plus "100,000 college professors in the U.S. alone who, according to Harvard researchers, agree that 'intelligent design IS a serious scientific alternative to the Darwinian theory of evolution.'" plus "570,000 medical doctors in the U.S., specialists in applied science, say God brought about or directly created humans." Enyart's "Honorable Mention: - 2.5 Million U.S. scientists and engineers believe in a personal God as reported by the New York Times in 1997..." [6]

Dr. Francis S. Collins is a leading geneticist and was head of the Human Genome Project. In his book, *The Language of God: A Scientist Presents Evidence for Belief,* he quotes Albert Einstein's "carefully chosen words" as saying, *"Science without religion is lame, religion without science is blind."* [7]

Another quote from Dr. Collins, *"Science is not threatened by God; it is enhanced. God is not threatened by science; He made it all possible. So let us together seek to reclaim the solid ground of an intellectually and spiritually satisfying synthesis of all great truths. That ancient motherland of reason and worship was never in danger of crumbling. It never will be. It beckons all sincere seekers of truth to come and take up residence there. Answer that call. Abandon the battlements. Our hopes, joys, and the future of our world depend on it."* [8]

In this book we expand substantially on the reasons why the Creator God of the Bible was formally and officially accepted into areas of our Governments at all levels in the four nations in our focus.

It is for these reasons, both scientific and historic, that students especially, have the inalienable right to be taught about the God of Their Governments.

We are in no way suggesting this be construed as a particular "religion," other than its becoming a "belief as a truth." It can replace other current "beliefs" such as evolution as the cause of life.

It is worthy of note that the majority of citizens in each of the four nations, claim to believe in God, according to census results noted at the end of our "God in Government" chapters.

If you think it through, you may agree that our Creator and Provider deserves our respect and appreciation for all the phenomenal, caring work He performs for each of us every second of every day.

This respect and appreciation is the reason for God's inclusion in our Governments in the first place.

Beyond His caring work, is the hugely beneficial advice He makes available for governments, teachers, students, business people, and all citizens.

It is time to make the change.

It will take many voices uniting to bring our Creator back to our classrooms.

The intent of the Atomic Biology Institute is to provide approved scientific reasons to do so.

If you believe this is a worthy cause, you can write or talk to your pertinent education leaders and/or you can sign a petition at www.atomicbiology.com **.**

End Notes and References

To make a significant difference in our education systems whereby Darwinisms and macro-evolution are eliminated as the taught cause of life, and the Creator God in our governments is the replacement, will take cooperation of many interested individuals and organizations.

Your comments would be appreciated. If interested, please contact us by email at admin@realityrandd.com or see our website at www.realityrandd.com .

Prologue

[1] Darwin, Charles, *On the Origin of Species by Means of Natural Selection*, 1st edition, 1859, p. 189, John Murray, London, Eng., available online from Darwin-online.org.uk.

Chapter 1: The Essentiality of a Super-Intelligent Force

[1] Darwin, Charles, *On the Origin of Species by Means of Natural Selection,* 1st edition, 1859, p. 189, John Murray, London, Eng., available online from Darwin-online.org.uk.

[2] Ibid., p. 280.

[3] Ibid., p. 186.

[4] Ibid., p. 189.

[5] Cohen, I. L., *Darwin was Wrong—A Study in Probabilities,* New Research Publications, New York, NY, 1984, p.5.

[6] Ibid., pp. 209–210.

[7] Reber, Paul, "What Is the Memory Capacity of the Human Brain," *Scientific American,* May/June, 2010, www.scientificamerican.com/article/what-is-the-memory-capacity/.

[8] Wedeen, Van, and Wald, L. L., in article "Secrets of the Brain" by Carl Zimmer, *National Geographic,* February, 2014, p. 34.

[9] Lichtman, Jeff in article "Secrets of the Brain" by Carl Zimmer in *National Geographic,* February, 2014, pp. 39, 43.

[10] Axe, Douglas, *Undeniable,* Harper One, New York, NY, 2016, p.259

[11] Diamond, Suzanne M., "Canada's Amazing Anti-Cancer Tea," *Common Ground,* December, 1999, pp. 18,19.

[12] Diamond, Suzanne M., "Humble Herbs Worth Their Weight in Gold," *Total Health Magazine*, American Wellness Network, Utah, USA, July/August 2008 issue.

[13] Diamond, Suzanne M., 2008, *Nature's Best Heart Medicine,* The Book Publishing Company, Summertown, TN. https://bookpubco.com/products/natures-best-heart-medicine .

Chapter 2: Our Phenomenal DNA

[1] "The Structure and Function of DNA," *Molecular Biology of the Cell*, 4th edition, National Center for Biotechnology Information, Bethesda, MA, www.ncbi.nlm.nih.gov/books/NBK26821/, accessed July 3, 2014.

[2] Gates, Bill, Myhrvold, Nathan and Rinearson, Peter, 1996, *The Road Ahead: Completely Revised and Up-To-Date,* Penguin Books, New York, NY, p. 228.

[3] Cohen, I. L., 1984, *Darwin was Wrong—A Study in Probabilities,* Research Publications, New York, NY, p. 5.

Chapter 3: Our Amazing Systems and Senses

[1] Ham, K., Varnum, P., and Mason, D. (Producers), Varnum, P. (Director), *Fearfully and Wonderfully Made,* DVD, Answers in Genesis Productions, Hebron, KY, 2007.

Chapter 4: Our Incredible Molecular Machines

[1] Sauvage, J-P., Stoddart, J.F., Feringa, B.L., 2016, www.nobelprize.org/nobel-prizes/chemistry/laureates/2016/press.html , accessed Nov.28, 2016.

[2] Tour, James, 2016, www.jmtour.com/about/research-information , accessed Nov. 28., 2016.

[3] Luskin, Casey, *Molecular Machines in the Cell,* Discovery Institute, 2010, www.discovery.org/a/14791 , accessed Nov. 28.,2016

Chapter 5: Moving Darwinism, Neo-Darwinism, and Macroevolution to the History Department

[1] Hoyle, Sir Fred and Wickramasinghe, Chandra *Evolution from Space,* J. M. Dent & Sons, London, Eng., 1981, p. 148.

[2] Meyer, Stephen C., *Signature in the Cell,* Harper Collins, New York, NY, 2009, p. 201.

[3] Darwin, Charles, *On the Origin of Species by Means of Natural Selection*, 1st edition, 1859, p. 189, John Murray, London, Eng., available online from Darwin-online.org.uk.

[4] Ibid., p. 186.

[5] Ibid., p. 189.

[6] Darwin, Charles, in a letter to Asa Gray (April 3, 1860), as cited in Norman MacBeth's *Darwin Retried: An appeal To Reason*, Gambit, Boston, MA, 1971, p. 101.

[7] *A Scientific Dissent from Darwinism*, www.dissentfromdarwin.com, accessed May 20, 2014.

[8] Ruloff, W., Sullivan, J., Logan Craft, (Producers), & Frankowski, N., (Director), *Expelled: No Intelligence Allowed*, DVD, Premise Media Corporation & Rampant Films, 2008.

[9] Luskin, C., "Bullies-r-us: how 'freethought oasis' threatened 'disruption' and pressured a college into cancelling intelligent design course," Dec. 10, 2013, *Evolution News and Views*, www.evolutionnews.org/2013/12/freethought-oasis-bullies079991.html, accessed May 20, 2014.

[10] Lewontin, R., Billions and billions of demons, *The New York Review of Books*, New York, NY, January 9, 1997, p. 31.

[11] Rogers, T., (Producer), & Woodyard, J. (Music Director), *Goodbye Evolution,* CD, Creation Studios, Canada, 2009.

[12] See Chapter 8 end notes.

[13] Darwin, Charles, *On the Origin of Species by Means of Natural Selection,* 1st edition, 1859, p. 280, John Murray, London, Eng., available online from Darwin-online.org.uk.

[14] Parker, Gary, in Paul S. Taylor's *Origins Answer Book*, fourth edition, Eden Productions, Mesa, Arizona, 1993, p. 103.

[15] Currie, Kenneth L., "Uniformity, Uniformitarianism, and the Foundations of Science" in Paul A. Zimmerman's *"Rock Strata and the Bible Record,"* Concordia Publishing House, St. Louis, MO, 1970, p. 40.

[16] Kemper, Gary, Kemper, Hallie, and Luskin, Casey, *Discovering Intelligent Design: A Journey Into The Scientific Evidence*, Discovery Institute Press, Seattle, WA, 2013, p. 221.

[17] Denton, Michael, *Evolution: A Theory in Crisis,* Burnett Books, London, Eng., 1985, p. 342.

[18] Cohen, I. L., *Darwin Was Wrong—A Study In Probabilities,* New Research Publications, New York, NY, 1984, p. 5.

[19] Hoyle, Sir Fred, and Wickramasinghe, Chandra, *Evolution From Space*, Aldine House, London, Eng., 1981, p. 149.

[20] Thaxton, Charles B., Bradley, Walter L., and Olsen, Roger L., *The Mystery Of Life's Origin: Reassessing Current Theories,* pp. 211–212, Philosophical Library, New York, NY, 1984, pp. 211.

[21] Grebe, John J., "DNA Complexity Points to Divine Design," *Science & Scripture,* vol. 3, no. 3, p. 20, Creation Science Research Center, San Diego, CA, 1973.

[22] Saunders, Peter and Ho, Mae-Wan, "Is Neo-Darwinism Falsifiable? And Does It Matter?", *Nature and System,* vol. 4, no. 4, December 1982, Tucson, AZ, p. 191.

[23] Ambrose, Edmund J., *The Nature and Origin of the Biological World,* John Wiley & Sons, New York, NY, 1982, p. 164.

[24] Hatsopoulous, G. N. and Gyftopoulos, E. P., *Deductive Quantum Thermodynamics,* Mono Book Corporation, Baltimore, MA, 1970, p. 78.

[25] Gish, Duane, "A Consistent Christian-Scientific View of the Origin of Life," *Creation Research Society Quarterly,* vol. 15, no. 4, March 1979, pp.186, 199.

[26] Smith, Wolfgang, *Teilhardism and the New Religion: A Thorough Analysis of the Teachings of Pierre Teilhard de Chardin*, Tan Books and Publishers, Rockford, IL, 1988, p. 8.

[27] Kemp, Tom, "A Fresh Look at the Fossil Record," *New Scientist,* vol. 108, no. 1485, December 1985, p. 66.

[28] Kitts, David B., "Paleontology and Evolutionary Theory," *Evolution,* vol. 28, September 1974, p. 467.

[29] Meyer, Stephen C., *Darwin's Doubt: The Explosive Origin Of Animal Life,* p.105, Harper-Collins Publishers, New York, NY, 2013, p. 105.

[30] Smith, J. Wolfgang, 1988, *Teilhardism and the New Religion: A Thorough Analysis of the Teachings of Pierre Teilhard de Chardin,* Tan Books and Publishers, Rockford, IL., p. 248.

[31] Denton, Michael, *Evolution: A Theory In Crisis,* Adler & Adler Publishers, Bethesda, MD, 1986, p. 342.

[32] Muggeridge, Malcolm, *Pascal Lectures,* University of Waterloo, Waterloo, ON, Canada, 1978.

[33] Gates, Bill, Myhrvold, Nathan, and Rinearson, Peter, *The Road Ahead: Completely Revised and Up-To-Date,* Penguin Books, New York, NY, 1996, p. 228.

[34] Sarfati, Jonathan, *Refuting Evolution, 4th edition*, Creation Ministries International, Eight Mile Plains, QLD, Australia, 2008, p. 121.

[35] Meyer, Stephen C., *Darwin's Doubt: The Explosive Origin Of Animal Life,* Harper-Collins Publishers, New York, NY, 2013, p. iv.

[36] Taylor, Paul S., *The Illustrated Origins Answer Book,* fourth edition, Eden Productions, Mesa, AZ, 1993, pp. 25–26.

[37] Cohen, I. L., *Darwin Was Wrong—A Study in Probabilities*, New Research Publications, Greendale, NY, 1984, pp. 209–210.

Chapter 8: A Fresh Introduction to the Scientific God of Our Nations

[1] References and notes relating to the enormous work performed for each of us just in constantly replacing our red blood cells:

Pallister, C. J., *Haematology: Biomedical Science Explained,* Butterworth-Heinemann, Burlington, MA, 1999. He states that an average 70 kg adult male produces *(or has produced for him)* about 2,300,000 red blood cells every second.

Tortora, G. J., *Principles of Anatomy and Physiology,* John Wiley & Sons, New York, NY, 2008. He states that there are approximately 280,000,000 molecules of hemoglobin per red blood cell.

Perutz, Max, *Science is Not a Quiet Life: Unraveling the Atomic Mechanism of Hemoglobin,* World Scientific, Hackensack, NJ, 1997. He states that each hemoglobin molecule contains approximately 10,000 atoms.

If you do the math, you will find that the number of atoms required to be sorted from our eaten food, then selected, counted, grasped, and assembled into new red blood cells and delivered into our bloodstream is approximately 2,300,000 x 280,000,000 x 10,000 = 6,440,000,000,000,000,000 (6,440 quadrillion) atoms every second of every day.

That approximate number is required for each average body every second of every day just for replacement red blood cells (based on a 70 kilo [154 lb.] male as average). We have conservatively used the figure of "over 4,900 quadrillion atoms per second" to include each of virtually all adults in the world.

Plus, logically, in the same second an even *greater* number of atoms (for roots, leaves, etc.) have to be found, sorted, selected, counted, grasped, assembled, and delivered from the soil, air, and water to make more food for each future second's food requirements for new red blood cell construction for each of us every second of every day.

Again, this is in addition to the phenomenal work of constantly sustaining, maintaining, repairing, and replacing the other approximately 80 trillion cells in each adult human body.

Can we possibly understand how much brilliant, reliable, trustworthy work and care this amounts to for each one of us?

Does God deserve our appreciation?

[2] Darwin, Charles, *On the Origin of Species by Means of Natural Selection,* 1st edition, 1859, p. 189, John Murray, London, Eng., available online from Darwin-online.org.uk.

Chapter 9: God in the Government of the U.S.A.

[1] "Declaration of Independence," *The Charters of Freedom,* The U.S. National Archives and Records Administration, College Park, MD, http://www.archives.gov/exhibits/charters/declaration.html, accessed March 1, 2014.

[2] "Transcript of Articles of Confederation (1777)," *Our Documents,* http://www.ourdocuments.gov/doc.php?doc=3&page=transcript, accessed March 1, 2014.

[3] "Transcript of Constitution of the US (1787)," *Our Documents,* http://www.ourdocuments.gov/doc.php?doc=9&page=transcript, accessed March 1, 2014.

[4] "U.S. Pledge of Allegiance to the Flag," http://www.publications.usa.gov/epublications/ourflag/pledge.html, accessed March 3, 2014.

[5] Courtesy of the Lillian Goldman Law Library's Avalon Project, Yale University Law School, http://avalon.law.yale.edu/subject_menus/inaug.asp, accessed March 3, 2014.

[6] "History of 'In God We Trust,'" U.S. Department of the Treasury, http://www.treasury.gov/about/education/Pages/in-god-we-trust.aspx, accessed March 3, 2014.

[7] "The U.S. National Anthem," U.S. Army Music, http://www.music.army.mil/music/nationalanthem/, accessed March 3, 2014.

[8] "The Court and Its Procedures," Supreme Court of the United States, Washington, DC, http://www.supremecourt.gov/about/procedures.aspx, accessed March 5, 2014.

[9 & 10] "Text for the Oaths of Offices for Supreme Court Judges," Office of the Curator, Supreme Court of the United States, Washington, DC, http://www.supremecourt.gov/about/oath/textoftheoathsofoffice2009.pdf, accessed March 4, 2014.

[11] "The Thomas Jefferson Building," *On These Walls: Inscriptions and Quotations in the Buildings of the Library of Congress,* U.S. Library of Congress, Washington, DC, http://www.loc.gov/loc/walls/jeffl.html, accessed June 16, 2014.

[12] "Thanksgiving in North America: From Local Harvests to National Holiday," Smithsonian, Washington, DC, http://www.si.edu/Encyclopedia_SI/nmah/thanks.htm, accessed March 5, 2014.

[13] Gallup Poll, May 5-8, 2011, results summarized in "More than 9 in 10 Americans Continue to Believe in God," June 3, 2011, Gallup, Inc., www.gallup.com/poll/147887/Americans-Continue-Believe-God.aspx, accessed Aug. 12, 2014.

Chapter 10: God in the Government of the U.K.

[1] "Coronation Oath Act, 1688," The National Archives, Government of the United Kingdom, http://www.legislation.gov.uk/aep/WillandMar/1/6, Richmond, Surrey, Eng., accessed March 6, 2014.

[2] "Coronation Oath, 2 June, 1953," The British Monarchy, http://www.royal.gov.uk/ImagescanBroadcasts/Historic%20speeches%20Oath2June1953.aspx, and%20broadcasts/CoronationOath2June1953.aspx , accessed March 6, 2014.

[3 & 4] "Oaths," Courts and Tribunal Judiciary, London, Eng., http://www.judiciary.gov.uk/about-the-judiciary/oaths/, accessed March 6, 2014.

[5] "Oath of Allegiance," U.K. Parliament, London, Eng., http://www.parliament.uk/site-information/glossary/oath-of-allegiance/, accessed March 8, 2014.

[6] "Oath of Allegiance," U.K. Parliament, London, Eng., http://www.publications.parliament.uk/search/results/?q=Oath+of+allegiance, accessed May 2, 2014.

[7] "Parliamentary Sovereignty," U.K. Parliament, London, Eng., http://www.parliament.uk/about/how/sovereignty/, accessed May 2, 2014.

[8] "National Anthem," The British Monarchy, http://www.royal.gov.uk/MonarchUK/Symbols/NationalAnthem.aspx, accessed March 7, 2014.

[9] "Processional Crosses", Westminster Abbey, London, Eng., http://www.westminster-abbey.org/worship/processional-crosses, accessed July 24, 2014.

[10] "Abbey Bells," Westminster Abbey, London, Eng., http://www.westminster-abbey.org/our-history/abbey-bells, accessed July 24, 2014.

[11] "Carola Morland," Westminster Abbey, London, Eng., http://www.westminster-abbey.org/our-history/people/carola-morland, accessed July 24, 2014.

[12] "Big Ben: The Clock Dials," UK Parliament, London, Eng., http://www.parliament.uk/about/living-heritage/building/palace/big-ben/building-clock-tower/clock-dials/, accessed July 25, 2014.

[13] "Appendix J: Prayers for the Parliament," *Companion to the Standing Orders and Guide to the Proceedings of the House of Lords,* 2013 edition, UK Parliament Publications, http://www.publications.parliament.uk/pa/ld/ldcomp/compso2013/Part3_14.htm, accessed Aug. 12, 2014.

[14] War Memorials Archive, Imperial War Museum, London, Eng., http://www.ukniwm.org.uk and various links, accessed June 13 2014.

[15] "Prayers," UK Parliament, London, Eng., http://www.parliament.uk/about/how/business/prayers, accessed June 14, 2014.

[16] "Religion by measures," Nomis, Durham, Eng., http://www.nomisweb.co.uk/census/2011/KS209EW/view/2092957703?cols=measures, accessed July 31, 2014.

[17] "Summary: Religious Group Demographics, 2011," The Scottish Government, Edinburgh, Scotland, http://www.scotland.gov.uk/Topics/People/Equality/Equalities/DataGrid/Religion/relPopMig, accessed July 31, 2014.

[18] "A note on the background to the religion and 'religion brought up in' questions in the Census, and their analysis in 2001 and 2011," Census Archives, Northern Ireland Statistics and Research Agency, http://www.nisra.gov.uk/archive/census/2011/Background_to_the_religion_question_2011.pdf, accessed July 31, 2014.

Chapter 11: God in the Government of Australia

[1] Dampier, W., *A Voyage to New Holland, an English Voyage of Discovery to the South Seas in 1699,* facsimile edition, 1981, accessible at Project Gutenberg, Salt Lake City, UT, www.gutenberg.org/files/15675/15675-h/15675-h.htm, accessed Aug.12, 2014.

[2] Scott, E., *The Life of Matthew Flinders,* Angus & Robertson, Sydney, AU, 1914, p.272.

[3] Clark, CMH., *A History of Australia Vol.1,* University Press, Melbourne, AU, 1962, p.80.

[4] Clark, CMH., *A History of Australia Vol.3,* University Press, Melbourne, AU, 1973, p.40.

[5] *Standing Orders, Legislative Assembly, Parliament of New South Wales*, p. 12, Parliament of New South Wales, Sydney, http://www.parliament.nsw.gov.au/Prod/la/precdent.nsf/0/0D813F110566E803CA2572A500059E36/$file/standing%20orders%202010.pdf, accessed Aug. 12, 2014 ,

[6] "Prayer and acknowledgement of country," *Chapter 8: Sittings, quorum, and adjournment of the Senate*, Powers, Practice, and Procedure, Parliament of Australia, Canberra, AU, http://www.aph.gov.au/About_Parliament/Senate/Powers_practice_n_procedures/aso/so050, accessed March 7, 2014.

[7] La Nauze, J., *Alfred Deakin, A Biography,* Angus & Robertson, Sydney, AU, 1979, pp.70, 79.

[8] McLennan, G., *Understanding Our Christian Heritage,* vol. 2, prayer 223, Christian History Research Institute, Orange, NSW, Australia, 1989, p. 80.

[9] "Commonwealth of Australia, Constitution Act," Powers, Practice, and Procedure, Parliament of Australia, Canberra, AU, www.aph.gov.au/About_Parliament/Senate/Powers_practice_n_procedures/Constitution/preamble, accessed March 8, 2014.

[10] "Commonwealth of Australia, Constitution Act," *Chapter V: The States*, Powers, Practice, and Procedure, Parliament of Australia, Canberra, AU, www.aph.gov.au/About_Parliament/Senate/Powers_practice_n_procedures/Constitution/~/~/~/~/link.aspx?_id=6ED2CAE61E7742A1B2C42F95D4C05252&_z=z, accessed March 8, 2014.

[11] McLennan, G., *Understanding Our Christian Heritage,* vol. 1, Christian History Research Institute, Orange, NSW, Australia, 1989, p. 49.

[12] Ibid., p. 51.

[13] "Tom Roberts' Big Picture," Parliament House Art Collection, Parliament of Australia, Canberra, AU, www.aph.gov.au/Visit_Parliament/Parliament_House_Art_Collection/Tom_Roberts_Big_Picture, accessed March 9, 2014.

[14] McLennan, G., *Understanding Our Christian Heritage,* vol. 2, Christian History Research Institute, Orange, NSW, Australia, 1989, pp. 81–82.

[15] "Recessional," 1897, notes by Mary Hamer, Jan. 24, 2008, The Kipling Society, Essex, Eng., www.kiplingsociety.co.uk/rg_recess1.htm, accessed Aug. 12, 2014.

[16] "Cultural Diversity in Australia — Reflecting a Nation: Stories from the 2011 Census," June 21, 2012, Australian Bureau of Statistics, Australia, www.abs.gov.au/ausstats/abs@.nsf/Lookup/2071.0main+features902012-2013, accessed Feb. 2, 2014.

Chapter 12: God in the Government of Canada

[1] "The London Conference December 1866 – March 1867," Canadian Confederation, Library and Archives Canada, Ottawa, ON, www.collectionscanada.gc.ca/confederation/023001-2700-e.html, accessed March 12, 2014.

[2] Reproduced with the permission of Rogers Communications Inc.

[3] "Constitution Act, 1982," Justice Laws, Government of Canada, www.laws-lois.justice.gc.ca/eng/CONST/page-15.html#docCont, accessed March 13, 2014.

[4] "Oaths of Allegiance Act," 1985, Justice Laws, Government of Canada, http://laws-lois.justice.gc.ca/eng/acts/O-1/page-1.html accessed Aug. 12, 2014.

[5] "National Anthem," The British Monarchy, London, Eng., www.royal.gov.uk/MonarchUK/Symbols/NationalAnthem.aspx, accessed July 9, 2014.

[6] "National Anthem: O Canada," Canadian Heritage, Government of Canada, Ottawa, ON, www.pch.gc.ca/eng/1359402373291/1359402467746, accessed July 10, 2014.

[7] "Thursday, 17th November, 1867," *Throne Speech*, Parliament of Canada, Ottawa, ON, www.parl.gc.ca/Parlinfo/Documents/ThroneSpeech/1-01-e.pdf, accessed July 12, 2014.

[8] "Speeches from the Throne and Motions for Address in Reply," Parliament of Canada, Ottawa, ON, www.parl.gc.ca/Parlinfo/compilations/parliament/ThroneSpeech.aspx?Language=E, accessed November 28, 2016.

[9] "Prayers, Daily Proceedings," *House of Commons Procedure and Practice*, 2nd edition, 2009, Parliament of Canada, Ottawa, ON, www.parl.gc.ca/procedure-book-livre/document.aspx?sbdid=af057bd0-f018-4fb4-bd75-4a2200729f05&sbpidx=2, accessed July 12, 2014.

[10] Veterans Affairs Canada, *Tomb of the Unknown Soldier,* www.veterans.gc.ca/eng/remembrance/memorials/canada/tomb-unknown-soldier , accessed June 30, 2014.

[11] War Monuments in Canada, *Vancouver: Victory Square Cenotaph,* www.cdli.ca/monuments/bc/victory.htm, accessed July 30, 2014.

[12] Matthews, J. S., *Remembrance Day 1944: Service of the Armed Forces and Citizens, Vancouver, Canada,* City of Vancouver Archives, Vancouver, BC, Canada, 1944.

[13] Veterans Affairs Canada, *In Flanders Fields,* www.veterans.gc.ca/eng/remembrance/history/first-world-war/mccrae, accessed July 30, 2014.

[14] Religion, Immigrant Status and Period of Immigration, *2011 National Household Survey: Data tables*, Statistics Canada, Ottawa, ON, www12.statcan.gc.ca/nhs-enm/2011/dp-pd/dt-td/Rp-eng.cfm?LANG=E&APATH=3&DETAIL=0&DIM=0&FL=A&FREE=0&GC=0&GID=0&GK=0&GRP=0&PID=105399&PRID=0&PTYPE=105277&S=0&SHOWALL=0&SUB=0&Temporal=2013&THEME=95&VID=0&VNAMEE=&VNAMEF=, accessed Aug. 12, 2014.

Brief Summary

[1] Cohen, I. L., *Darwin was Wrong—A Study in Probabilities,* New Research Publications, New York, NY, 1984, p.5.

[2] Ibid., pp. 209–210.

[3] References and notes relating to the enormous work performed for each of us just in constantly replacing our red blood cells:

Pallister, C. J., *Haematology: Biomedical Science Explained,* Butterworth-Heinemann, Burlington, MA, 1999. He states that an average 70 kg adult male produces *(or has produced for him)* about 2,300,000 red blood cells every second.

Tortora, G. J., *Principles of Anatomy and Physiology,* John Wiley & Sons, New York, NY, 2008. He states that there are approximately 280,000,000 molecules of hemoglobin per red blood cell.

Perutz, Max, *Science is Not a Quiet Life: Unraveling the Atomic Mechanism of Hemoglobin,* World Scientific, Hackensack, NJ, 1997. He states that each hemoglobin molecule contains approximately 10,000 atoms.

If you do the math, you will find that the number of atoms required to be sorted from our eaten food, then selected, counted, grasped, and assembled into new red blood cells and delivered into our bloodstream is approximately 2,300,000 x 280,000,000 x 10,000 = 6,440,000,000,000,000,000 (6,440 quadrillion) atoms every second of every day.

That approximate number is required for each average body every second of every day just for replacement red blood cells (based on a 70 kilo [154 lb.] male as average). We have conservatively used the figure of "over 4,900 quadrillion atoms per second" to include each of virtually all adults in the world.

Plus, logically, in the same second an even *greater* number of atoms (for food roots, leaves, etc.) have to be found, sorted, selected, counted, grasped, assembled, and delivered from the soil, air, and water to make more food for each future second's food requirements for new red blood cell construction for each of us every second of every day.

Again, this is in addition to the phenomenal work of constantly sustaining, maintaining, repairing, and replacing the other approximately 80 trillion cells in each adult human body.

Can we possibly understand how much brilliant, reliable, trustworthy work and care this amounts to for each one of us?

Does God deserve our appreciation?

[4] Darwin, Charles, *On the Origin of Species by Means of Natural Selection, or The Preservation of Favoured Races in the Struggle for Life*, 1st edition, 1859, p. 280, John Murray, London, Eng., available online from Darwin-online.org.uk.

[5] Parker, Gary, in Paul S. Taylor's *Origins Answer Book*, fourth edition, Eden Productions, Mesa, Arizona, 1993, p. 103.

[6] Enyart, Bob, *RSR's List of Scholars Doubting Darwin,* found at http://kgov.com/scientists-doubting-Darwin accessed March 6, 2017.

[7] Collins, Francis S., *The Language of God,* Free Press div. of Simon & Schuster, New York, NY, 2006, p.228.

[8] Ibid., pp.233-234

Credits and Permissions

The inclusion of any quotations, charts, figures, photographs, or other types of images in this book should not be considered as endorsements of the contents of this book by the copyright holders of those quotes or images.

Acknowledgements

First, we – at Reality R&D – must acknowledge our Creator God who provides virtually all people with phenomenal blessings including our marvelous, computer-like brains to logically and progressively determine how things work.

Then we must acknowledge the thousands of scientists who have collectively built upon each other's research and findings to gradually assemble the vast amount of knowledge and understanding we now enjoy and can employ in the betterment of our lives and society. Many have been quoted herein.

A special thanks goes to Dr. Graham McLennan for being the first historian to accept the task of writing a chapter regarding God's involvement in a national government, Australia in this case.

Then, I would like to thank the many friends and assistants who have been a great encouragement in the production of this primer and related textbooks in process.

Last but most, I thank my dear wife, Bonnie, for her patience, contributions, suggestions, proof-reading, and moral support over the many years of effort that have gone into this project thus far.

About the Authors

Ranson Williams/Worldwide Photos.

Dr. Graham McLennan is the author of Chapter 11: God in the Government of Australia. He received his Degree in Dentistry from Sydney University, where he also became a Christian. He served as a Captain in the Australian Army; received the Defence Medal and the National Serviceman's Medal.

Graham and his wife, Pam, founded the National Alliance of Christian Leaders (NACL) in 1973. Later, they founded Christian History Research Institute. He was on the executive of the National Gathering of 1988 and the Bicentennial of Christian Education in 1993. He also initiated the National Christian Heritage Sunday celebrations in 2012 and received the Presidential Medal from the President of Vanuatu for "services to the nation."

In addition to being a dental surgeon and tutor at the nearby University Dental School, Graham is convener of the NACL; team member for the Australian Christian Values Institute and National Day of Prayer and Fasting; founding Chairman of Rhema FM 103.5, and Orange Christian School. He is also a founding director of UCB's Vision FM, and of the Australian Christian Lobby and has served on many other national and international Christian boards and charities, as well as authoring a number of Christian books and articles.

Why did Dr. McLennan want to be a part of this project?

"Every time I look at a beautiful scene, a child, a magnificent flower or bird, I think surely this hasn't been produced from primordial soup on the basis of time and chance.

Yet, especially through our media, education, and government policy, secularists have deceived many into believing this lie, the consequences often being despair devoid of purpose and meaning.

A sad indictment on our society when we can see God providentially moving in history – this is HisStory in our lives and nations, that we must continue to teach our children. Psalm 78:4

This is why we support further research through the Atomic Biology Institute."

Dr. Graham McLennan
National Alliance of Christian Leaders
Australia

RealityRandD.com

Thomas Rogers is an independent researcher, president of Reality Research and Development Inc., and the Atomic Biology Institute, as well as other companies. He has studied at three universities and two specialty institutes. His work background is in engineering, research, construction, international manufacturing, and exploration. His interest in sciences began in elementary school and has been ongoing for decades beyond university, at all times seeking practical and significant applications.

The education and experiences in these fields helped him in researching and understanding the super-intelligent physical work required to design, construct, sustain, and maintain living entities. The basics of God's brilliant work in building humans and their foods, using super-intelligence, vision, dexterity, precision, speed, and enormous, reliable care for each one of us, are detailed herein. He concurs with I. L. Cohen that claiming evolution is the cause of life, is probably the worst mistake ever made in science.

Tom has been a voluntary director and president of various community organizations, including the British Properties & Area Homeowners Association, the Greater Vancouver Apartment Owners Association, the West Vancouver – Howe Sound Social Credit Constituency Association, and is a member of the Salvation Army. He has memberships in the American Scientific Affiliation, the Canadian Scientific and Christian Affiliation, the Discovery Institute/Center for Science and Culture, the Creation Science Association of British Columbia, the Christian Scientific Society, and the American Association for the Advancement of Science.

Index: (See also - Glossary

H

I

J

L

M

O

P

Like To Be Involved?

If you believe this information is significant for our society and should be shared, please send a note to your friendly contacts. It will take many believers and respecters of truth, to help bring the God of our nations back to our classrooms. This is a huge task that will take the endorsement of many citizens.

Please see our website www.realityrandd.com .

Coming Textbooks and Contact Information

There are many basic aspects of Atomic Biology presented in this primer, therefore, it can be used as a textbook for introducing this new subject. It provides details of how we are made and cared for.

The further development of this science in replacing Darwinisms as the cause of life, is an ongoing project by the Atomic Biology Institute and Reality Research & Development Inc.

A more detailed textbook is initiated for deeper study by college and university students interested in the application of this new science.

We seek reputable scientists to write some of the chapters. There is a compensation contract for each writer. If interested and/or if you have comments or interest in supporting this project of bringing God back to our classrooms, please contact us:

By Email:
admin@realityrandd.com
By Facebook via our FB page:
https://www.facebook.com/DarwinsReplacement/

By Twitter Direct Messaging (DM):
https://twitter.com/DarwinsReplace

By Contact Form:
http://www.atomicbiology.com/contact-us/

Made in the USA
Lexington, KY
23 April 2018